Whole Life Sustainability

Ian Ellingham and William Fawcett

RIBA  **Publishing**

© Ian Ellingham and William Fawcett, 2013

Published by RIBA Publishing, 15 Bonhill Street, London EC2P 2EA

ISBN 978 1 85946 450 2

Stock code 77534

The rights of Ian Ellingham and William Fawcett to be identified as the Authors of this Work have been asserted in accordance with the Copyright, Design and Patents Act 1988.

British Library Cataloguing in Publications Data
A catalogue record for this book is available from the British Library.

Commissioning Editor: James Thompson
Project Editor: Neil O'Regan
Designed and typeset by Kneath Associates
Printed and bound by Polestar Wheatons

While every effort has been made to check the accuracy and quality of the information given in this publication, neither the Author nor the Publisher accept any responsibility for the subsequent use of this information, for any errors or omissions that it may contain, or for any misunderstandings arising from it.

RIBA Publishing is part of RIBA Enterprises Ltd.
www.ribaenterprises.com

Whole Life Sustainability

Ian Ellingham and William Fawcett

RIBA ╫ Publishing

Contents

The Authors

Ian Ellingham BArch MBA MPhil PhD FRAIC trained as an architect at Carleton University in Canada. He studied for the MBA degree at the University of Western Ontario, MPhil at the Department of Land Economy in the University of Cambridge, and PhD at the Department of Architecture in the University of Cambridge. His PhD thesis was 'Consumer influences on the life-cycle of housing'. After twenty years in the development industry he now focuses on research, writing and teaching. He has been an associate of Cambridge Architectural Research Limited since 1999.

William Fawcett MA DipArch PhD RIBA trained as an architect at the Department of Architecture in the University of Cambridge, where he went on to PhD research on 'A mathematical approach to adaptability in buildings'. As well as working in architectural practice, he has held a lectureship at the University of Hong Kong and the Chadwick Fellowship in Architecture at Pembroke College, Cambridge. He has been a director of Cambridge Architectural Research Limited since the company was founded in 1987.

Drs. Ellingham and Fawcett have undertaken numerous research studies for both government and industry, and have written widely on issues in the built environment.

Preface

This book is published in 2013 but - as might be expected - has its roots much further back. A formative influence was a UK government funded research project called *Evaluating Sustainable Investment in an Ageing Building Stock*, begun in 1998 and led by Cambridge Architectural Research Limited. The primary finding was unexpected: that sustainability depends on good decision-making skills. Without appropriate tools, decision-makers are essentially reduced to throwing collections of possible solutions at problems, and hoping something might stick. The heroes of this first research project were experienced older practitioners who could assess a set of complex issues and come up with sound decisions. The problem was that they couldn't explain how they did it - their wisdom relied on internalised experience built up over decades of trial and error. Acquiring decision-making skills in this way has two problems: it takes too long and involves too many mistakes. The alternative is to rely on good decision-making tools. Our first book, *New Generation Whole-life Costing* (2006), focused on a particular decision-making tool, whereas this book offers a broader overview.

The financial chaos and prolonged recession following 2008 have underlined the need to use all resources more carefully. Decisions are truly difficult. Competing claims on resources often come down to the trade-off between alternatives with present benefits but future costs, as against others with present costs but future benefits. When applied to investment in large, resource-hungry, long-lived assets such as buildings, the objective is to strike a balance between under-investment (building to a low standard and regretting it later) and over-investment (expenditure of resources without gaining a commensurate return). The difficulty of finding the balance is amplified because the future is never certain. Dealing with future uncertainty could be the most important challenge for people who aspire to create and manage sustainable buildings.

This book looks at various ways of dealing with the issues of sustainability. It presents methods developed in other fields, and suggests ways in which the decision-maker can apply them within the context of buildings and land. In particular, this book is about the quest for balanced decision-making. The authors accept that most decisions will not be made using detailed calculations, often because data is either too expensive or actually impossible to collect. Consequently they have focused on concepts rather than formulae. Many of the decision-making techniques are drawn from economics and management, because these disciplines have been considering how to approach such problems ever since they were founded.

Research in the area is ongoing. This book takes account of current work on the EC-funded CILECCTA project (www.cileccta.eu) in which Cambridge Architectural Research Limited is a partner. This pan-European project is developing ideas and software that will enhance sustainable analysis and decision-making in the construction industry.

While bearing the imprint of RIBA Publishing, the book is written for a wider audience than just architects. Anyone involved in the creation or management of buildings should find useful ideas and be stimulated to perform better at the next project meeting.

Sustainability, Ubiquitous Sustainability

1

1.1 Lost For Choice

You want sustainability, your clients want sustainability, and practically all building products claim to be sustainable. Since we all agree, shouldn't sustainable design be easy? Well, no – how do you choose a route through the landscape of alternative design decisions when every signpost says 'Sustainability this way'?

To figure out what's *really* sustainable on a given project and for a given client, designers have to ask searching questions – of themselves, of their clients, of product manufacturers and of their buildings. This book offers suggestions about what to ask. Unfortunately, some of the questions may go unanswered, and some of the answers may be unconvincing. So designers often have to work out the answers for themselves, by pulling apart the bundle of issues that underlie the ubiquitous label 'sustainability'.

There's no doubt that sustainability is a very powerful idea, with enormously wide – perhaps unlimited – relevance. But this makes sustainable decision-making all the harder. Compare it to another very powerful and widely applicable concept:

fairness. Psychologists tell us that the sense of fairness or justice can be an even more powerful determinant of behaviour in humans and other animals than self-interest. People often voluntarily go to self-destructive lengths to correct a perceived unfairness. The trouble is that the same concept of fairness is used by other people to come to diametrically opposing, but equally strongly held, conclusions – keeping the lawyers (and arms salesmen) busy, unfortunately.

Let's hope that architects, in their limited sphere of activity, are dealing with a more tractable problem. We don't want Courts of Sustainability alongside Courts of Justice. For the architect the problem can be stated simply: *how can we create a more sustainable built environment*? Of course, it's tempting to take a slightly different view, and end up with an intractable problem such as *how can we save the planet*? We're sticking to the first version, in the belief that it makes it easier for architects to be able to make a modest but worthwhile contribution to the Big Picture.

1.1.1 How come everything is sustainable?

Sustainability must be one of today's most overused words. In every architectural journal or website we see numerous product and service providers trumpeting 'sustainability', 'green-ness' or 'eco-something'. While there is more to sustainability than other popular marketing slogans like 'new', 'healthy', 'all-natural' or 'wrinkle removing', caution should be observed and claims probed. For instance, we would be sceptical if we found electric patio heaters being promoted as sustainable because the sales manager cycles to work. (We invented that example, but you get the idea.) Often it comes down to selective reporting: practically every service or product has some feature that is genuinely sustainable, but positives have to be balanced against negatives, not viewed in isolation. Negative claims are equally open to scrutiny; should we be convinced by the proposition that natural stone is a finite resource that is being depleted and is therefore unsustainable? (That's a real example, not one we invented.) It is – possibly – literally true, and yet … surely absurd.

1.1.2 Wishful thinking

Over-enthusiastic marketing efforts make it difficult for architects to make rational decisions. An effective and widely used marketing strategy relies on potential consumers' longing for a better world, or even a better self. It exploits the 'self-concept': there is an actual condition (how people perceive themselves) and an ideal condition (how people would like to perceive themselves).[1] Marketing feeds on consumers' dissatisfaction with their current condition and engenders the belief that they could approach their ideal by buying some product or service. Advertising photos of building interiors, for example, are often shown with few furnishings and certainly no clutter – several white chairs, and perhaps a white carpet. Many people see this clean and serene world as more in keeping with their ideal self than the reality in which they actually exist – full of kids with toys, cats with claws, messy dogs, fuel bills, tax demands … so why not close the gap between the real and the ideal by purchasing something in the seductive image – say, a white leather and chrome Barcelona chair?

With sustainability, the marketing people have found a wonderful tool to encourage consumption. People are encouraged to feel that purchasing eco-products helps with the progression towards some ideal condition, and that if they do not make the purchase they should feel personally responsible for bad things happening to the planet.

We can't really blame the marketing people. They're paid to present a product or service in a way that will engage potential consumers most effectively: advertisers' promotion of sustainability is a direct response to consumers' longing for sustainability. It's equivalent to paying a barrister to plead for a defendant in the way that will most effectively engage the sympathy of the jury – however, the jury also hears from another barrister who paints the defendant in the blackest of terms … and then forms its own independent view.

When it comes to design decisions, architects are the jury. They should be aware of their own vulnerability to flattery and wishful thinking, aim to acquire insights and techniques, and – most important of all – think for themselves.

1.2 What Exactly is Sustainability?

Effective decision-making in the pursuit of sustainability has to rest on a set of solid, well-established principles. Otherwise decision-making is capricious and arbitrary – and may be risky. An American lawyer raised the alarm, observing that sustainability has 'fuzzy outlines', with 'green' being even worse.[2] He warned about potential liability issues for architects who make promises in such terms. There's an echo of the hot water that the airlines operating Concorde got into in the 1970s when they claimed that supersonic air travel eliminated jet lag. They were challenged in the American courts and resorted to the argument that 'jet lag' was a meaningless concept, so their claim couldn't be false; however, they still withdrew the claim. Let's not go down this route with sustainability.

People have different interpretations of what sustainability means, and therefore what courses of action might be considered sustainable. The development of a building involves, among other contributing factors, clients, designers, financial sources, builders, specialist consultants, suppliers, government bodies and facilities managers. Without a group consensus, or at least an acknowledgement of different attitudes, there is a risk of misunderstanding or even conflict over what ought to be a shared objective.

Many people focus on one aspect of sustainability, either choosing to ignore or being unaware of other aspects. Let's try to get an overview.

One might start with dictionary definitions. The *Oxford English Dictionary* defines three uses of the word **sustainable**: 1) Capable of being borne or endured; supportable, bearable. 2) Capable of being upheld or defended; maintainable. 3) Capable of being maintained at a certain rate or level.

The key point is that some present state of affairs can be sustained into the future – that the person, the structure, the resource supply and, very importantly, the rate of economic growth are not going to collapse.

Many discussions of sustainable development trace the concept to *Our Common Future*, the 1987 report by the United Nations World Commission on Environment and Development, usually referred to as the Brundtland Report. The term had been used earlier, notably in *World Conservation Strategy*, the 1980 report of the International Union for Conservation of Nature. It is a significant undertaking to read the full 374 pages of the Brundtland Report, but it puts the issues of the built environment into a bigger context. The headline definition cannot adequately capture the full complexity of sustainability, but it encapsulates the essence:

> *Sustainable development is development that meets the needs of the present without compromising the ability of future generations to meet their own needs. It contains within it two key concepts:*
> - *the concept of 'needs', in particular the essential needs of the world's poor, to which overriding priority should be given; and*
> - *the idea of 'limitations' imposed by the state of technology and social organization on the environment's ability to meet present and future needs.[3]*

This definition does not focus on the natural environment. The report links social, environmental and economic issues in the context of development – with development meaning the improvement of human welfare. In particular, consider:

> *The environment does not exist as a sphere separate from human actions, ambitions, and needs, and attempts to defend it in isolation from human concerns have given the very word 'environment' a connotation of naiveté in some political circles.[4]*

The Brundtland approach is that human development and environmental concerns should be complementary and compatible.

The Brundtland Report is not definitive. Groups that are highly focused on the natural environment see the definitions as being too human-centric, while industry often sees them as too focused on the environment. Many in the developing world view the document as representing a set of values being imposed by the wealthy that will limit opportunities for the less affluent. Others who are more optimistic about technical and economic progress see many sustainability concepts as irrelevant and potentially damaging to the world's material well-being.

A great deal of new knowledge has been accumulated since 1987. Nevertheless, it is worth reiterating some of the Brundtland Report's points, and considering them in relation to people who want to create and manage a sustainable built environment.

1.2.1 People-focused view

The Brundtland view of sustainability is focused on people. It does not mandate the preservation of natural environments, other forms of life or natural resources without regard to people or cost. It sees economic activity as an integral part of human life, but it does view military expenditure as a threat to continued human well-being.

1.2.2 Equity within and between generations

Sustainability embeds the concept of equity and justice both within and between generations. In our own time needs are relative, not absolute, being defined differently from place to place and from person to person. Brundtland says that the needs of the poor (e.g. food and basic healthcare) should take precedence over the needs of the rich (e.g. the latest clothing fashions).

Inevitably the situation is more complex when comparing the needs of people today with people in the future. The further one looks into the future, the greater the uncertainty about the needs of future generations. Consider some resources that were once vital but are no longer quite as important: coal for domestic heating, horses for transport, timber for ships' masts, whale-oil for lighting, asbestos for insulation and shellac for 78 rpm records. Can we realistically predict which of today's needs will evaporate and what new ones will emerge? The success of any sustainability-driven initiative taken today will not be determined by us, but by our intended beneficiaries in the future.

1.2.3 Technological progress

The world is changed by technology, often for the better.[5] Human capabilities cannot be regarded as static – continuing technological progress is likely to contribute to future well-being. In buildings, we have seen many inventions contribute to the occupants' well-being, including chimneys, plumbing, damp-proof courses, cheap glass, fibreglass insulation and artificial lighting. If the Georgians had decided to close down technical progress in the 18th century we would be worse off today; and if we close down technical progress today we will undoubtedly penalise future generations.

1.2.4 Economic development

Economic growth should not be taken for granted.[6] Among the criteria for human development, one of the most important is that the world's economy remains productive and forward-moving. Although there are stresses for people living in prosperity, it is incomparably worse to live in poverty. Economic shocks that throw large numbers of people into poverty should be seen as human tragedies, not dismissed as technical aberrations.

1.2.5 Environmental factors

Sustainability is not restricted to environmental or green factors, but is much broader. Rather unfortunately, there is a widespread perception that equates sustainability with environmental responsibility – so that 'sustainable' means no more than 'green'. This is a very incomplete view. It is inadequate even for the limited problem of sustainable decision-making in architecture.

1.3 Three Components of Sustainability

A professional involved in construction might think that sustainability as it applies in the built environment should focus on environmental issues, but the reality is that it will be difficult to make balanced, insightful decisions if one focuses only on greenness. Buildings inevitably consume resources, but in exchange they can improve the lot of humanity.

The idea of multidimensional assessment, sometimes referred to as a 'triple bottom line', has been around for some time, referring to social, economic and environmental dimensions of sustainability. This concept suggests that when policies, projects or actions are being evaluated, or alternatives being compared, they should be assessed with respect to all three dimensions. It makes sense, but is extraordinarily difficult to apply because the three dimensions are measured in such different ways. How do you combine the three dimensions for one project where, for example, social benefit comes at an environmental cost (the construction of a school will inevitably consume resources), or when environmental benefit comes at a social cost (closing down a polluting factory which makes its workforce redundant)? Or, even more difficult, how do you compare alternatives A and B, if A performs better in some dimensions but B is better in others?

Clear thinking and good judgement are in order. Take the case of air travel and its terrible carbon footprint. The environmental impact in terms of CO_2 emissions per kilometre is more or less constant regardless of the purpose of the flight (although it also depends on the age of the plane and certain other characteristics), but the social dimension depends on the purpose of the journey. A flight to enable a research study of earthquake-induced building collapse, which might reduce casualties in future events, is very different from flying to a drunken weekend in Ibiza. Arguably, the former has a significant benefit to many people, whereas the latter has a dubious benefit to a small number (hopefully) of individuals. A purely environmental view of sustainability would treat them both as identical and damaging, but a broader view that considers both social and environmental dimensions would distinguish between the research trip that contributes to human well-being and the hedonistic jaunt that detracts from sustainability (note that value judgements cannot be avoided).

In construction projects the three dimensions are practically always in play. Few buildings are put up just for the pleasure of building. They serve social (house, school, hospital) or economic (factory, office, warehouse) objectives – very often a mixture of both. It does not make much sense to ask whether a building is sustainable without knowing what it is for – but is it the architect's job to query the client's motives? Lecturing on factory design in 1927, the architect-engineer Sir Owen Williams offered his opinions:

'What is a Factory?' I take it we are only considering the 'buildings' of factories. I would define it as a place protected from wind and weather where things, mostly unnecessary, are made most efficiently. It is always dangerous to be curious as to the 'why' and 'wherefore' of the article to be manufactured. The result of any such investigation may be somewhat depressing to enthusiasm. It is enough that the manufacture must be most efficient.
Sir Owen Williams, 'Factories'[7]

In similarly dangerous territory, should one or should one not be sceptical if commissioned to design a 1,000 square metre sustainable house, with helipad, for a client who already owns two or three other houses? Could even the most environmentally friendly materials make such a project sustainable?

Most people in the construction industry are distanced from the set of social and economic decisions that create the projects on which they work. It is this that pushes the sustainability focus on to the environmental aspects of the project, over which they have more influence. Each person must decide whether to take Sir Owen Williams' extreme approach of detachment from the ultimate purpose of the building, or to seek a broadening of the vision. The creation of a building involves many big and small decisions, and each one may provide the opportunity for a triple bottom line approach to sustainability. Is it better to specify windows from an established multinational manufacturer or to go to a start-up company in a regeneration area? Dame Jane Drew described such a case, when hundreds of manual labourers worked on Le Corbusier's Indian buildings that were meant to be symbols of modernity: if they had brought in modern machinery, where would the labourers have found work?

It is worth thinking about the definition of sustainability recently proposed by the head of a major London-based commercial property company. He defined it as any collection of attributes that would enable a building to survive and experience increasing rents and yields over the longer term. The definition allows for, but is much broader than, environmental aspects of sustainability. It says that a sustainable commercial building will continue to attract tenants who want to occupy it and are able pay a commercial rent, and it will therefore achieve the social and economic objectives that led to its construction. It assumes that over a protracted period the market will determine the success or failure of the building, and will give rise to a balanced assessment with respect to social, economic and environmental performance. From the perspective of a practical creator of buildings, this overall performance-based concept might be useful. Lutzendorf and Lorenz (2005) also connected this type of definition to sustainability, and how environmental and social benefits can be expressed in terms of rents and capital values. Rather than attempting to incorporate a grab-bag collection of 'green' products in a project, this approach suggests that observable and quantifiable outcomes might be used to verify a successful, sustainable building, and lead to a better understanding of the issue. Thus, a sustainable building should be capable of addressing the ongoing expectations of owners and tenants – who are in turn attempting to meet the demands of future generations of customers, employees and other users, who will bid increasing amounts for the services offered by the building.

It is unfamiliar to see sustainability expressed in terms of commercial rents and yields, but a sustainable building should, in the first instance, be durable – indeed, 'sustainability' translates into French as *durabilité*. If a commercial building can no longer be let and is demolished after a short useful life, the resources that went into its construction are wasted, and it must be regarded as unsustainable, however many green features it incorporated. This doesn't just apply to commercial buildings. Greenwich District Hospital was a flagship social project of the early 1970s, built at high cost and following the latest ideas in flexible hospital planning, but it closed in 2003 and has since been demolished. Its inability to achieve its social objectives represented a sustainability failure.

1.4 Capital and Income

The word 'capital' often appears in connection with sustainability. It's based on an analogy with traditional businesses, in which capital assets are the machines, tools, factories, and other physical things that firms rely on to generate income. These capital assets are typical of manufacturing companies, but companies offering services, including architectural and engineering practices, also have capital assets, such as techniques, patents or skills embodied in the workforce.

Putting a value on capital assets is tricky. Every company's annual report includes a statement of the value of its capital assets, following accounting guidelines that aim to make valuation precise, factual and objective, so that one organisation can be compared with another. But the value put on assets may not reflect their productive worth. A very useful piece of equipment may be valued at the same amount as the one everyone regrets having purchased and is rusting in a back shed awaiting disposal. Accounting reports emphasise tangible assets, such as buildings, machinery and transportation equipment, often overlooking or undervaluing intangible assets, including intellectual property, patents, trademarks and managerial capability. When the Canadian telecom giant Nortel Networks Corporation failed, its patents were sold for $4.5 billion, far in excess of the tangible asset valuation in the company's financial statements.

What is meant by capital in the sustainability debate? The pioneering environmental economist David Pearce (1941–2005) distinguished three types of capital: 'natural capital', 'physical capital' and 'human capital'.[8] He argued that a combination of all three is required to support human well-being today and for future generations.

- **Natural capital** embraces the natural world and its ecology, and all sources of energy and materials that can be obtained from the natural environment to support human activities.

- **Physical capital** consists of everything tangible that people have created over time, including manufacturing equipment, buildings and infrastructure such as roads and airports, as well as works of art and other items of cultural heritage. Physical capital encompasses more than steel mills and sewers, embracing everything man-made that has value in different settings and cultures.

▓ **Human capital** is made up of the knowledge, skills and competencies and other attributes embodied in living people that are relevant to human activities. It embraces the sum of human knowledge and skills, individually and collectively. It is difficult to measure – a standard approach is to count years of education and levels of qualification in a population, but it covers more than just academic achievement; it also includes entrepreneurial ability and the ability to cooperate productively.

Pearce contrasted two views of sustainability. The strong view of sustainability is that each generation must pass on to the next an undiminished stock of natural capital. The weak view of sustainability is that each generation must pass on to the next an undiminished aggregate stock of natural, physical and human capital. The weak view implies that it is possible to substitute one form of capital for another. Thus, for example, the extraction and consumption of fossil fuel is unsustainable according to the strong view because it reduces natural capital, but it could be sustainable according to the weak view if the loss of fossilised carbon is more than balanced by an increase in physical or human capital, for example in the form of a hydroelectric dam.

Architects and engineers are more or less obliged to adopt a weak view of sustainability, because their efforts consume natural capital in order to create physical (and perhaps some human) capital. Most architectural debate about sustainability has been directed towards reducing the reliance on natural capital; much less attention has been directed towards buildings' potential for adding to the overall stock of capital assets. However, physical capital does form a significant resource, and what we build will be used by future generations in the same way that we use the buildings and infrastructure our predecessors left us.

The principles set out in Pearce's model of capital and sustainability are convincing, but it is difficult to estimate the relative values of the different kinds of capital. Nevertheless, trade-offs are made. Should the conservation of bats and newts (natural capital) take precedence over new construction (physical capital); should the conservation of architectural heritage (one type of physical capital) take precedence over new construction (another type of physical capital)? Every time people fly to a conference on environmental conservation they must believe that the gain in human capital due to knowledge exchange and opinion-forming outweighs the loss of natural capital due to burning aviation fuel.

1.5 Income and Consumption

In the business world, a company's stock of capital assets is important but reveals very little about its performance. What's more important is how the capital is used to generate a flow of income to the company, its employees, its shareholders and – through taxation – to wider society. The frequent failure of large corporations shows that having a large stock of capital assets can be meaningless without the ability to put them to productive use.

Similarly, concern with human well-being should have regard for the outcome
– how people live – rather than focusing on the stock of capital assets at their
disposal. We tend to condemn the folly of a miser who hoards goods and money
but lives an impoverished life. And few would envy traditional Highland crofters
living in the magnificent and unspoiled natural landscape of the glens. Despite an
abundance of natural capital their lives were cruelly hard.

How people live is related to their consumption of goods and services – food,
clothing, housing, fuel, healthcare, transport, communications, education, culture
and entertainment all contribute to quality of life. Talking about consumption
may seem overly materialistic, but what really undermines the well-being of the
poor and deprived parts of the human population is under-consumption of basic
goods and services. Over-consumption in parts of the developed world may be
reprehensible, but it does not make all consumption wrong.

Sustainability requires that there should be an undiminished level of well-being
over time, although the nature of that well-being might change. In fact, if the well-
being of future generations is the ultimate objective, support of consumption could
be considered a primary goal, with capital assets playing a supporting role. As John
Maynard Keynes said, 'Consumption – to repeat the obvious – is the sole end and
object of all economic activity.'[9]

1.6 Sustainability and Innovation

It is necessary to recognise the importance of innovation as a driver of human
well-being, and so a key part of sustainability. We should not assume that the
methods of today will be the methods of the future. Many improvements in
human well-being directly or indirectly result from innovation. If one reflects on
what has made the difference between the way people lived two hundred years
ago and now, significant factors would include improved medical care, public
health and advances in agricultural technology – all the results of innovation. The
Brundtland Report points to the promise of new capabilities: 'technology and social
organization can be both managed and improved to make way for a new era of
economic growth'.[10]

In the developed world we have become accustomed to the benefits that result
from innovation, yet advances in knowledge and technology cannot be taken
for granted. Historically, the wealth of societies and well-being of their members
does not always continue to increase: there are many examples of societies
that have flourished and then declined. The decline of each civilisation and the
disappearance of its intellectual life and innovation may have different reasons.
However, such declines do point to the need to be concerned about the forces that
might bring an end to our own era of progress.

Immense amounts have been written about innovation – what causes it, and its
benefits. Some warn against the dangers inherent in bureaucratic environments.

Milne (2011b) noted that procurement policies increasingly focus on the volume of past experience, rather than 'ideas and capabilities'.

Pearce reflected on global warming: 'Perhaps the mistake in the debate has been to get too hung up about warming itself, rather than focusing on the need to get technology moving.'[11] Could a too-strong focus on the environment in itself interfere with innovation processes, thereby undermining crucial elements in the long-term improvement of human well-being?

For the practitioner attempting to make decisions, the potential of innovation changes things. It tends to work against pessimistic forecasts, because the presence of innovation undermines simple projections based on existing trends and conditions. The appearance of new products, techniques and methods means that the decision-maker must expect that something new might change everything. The mere possibility of that happening may have an influence on deciding the best solution. Accordingly, alternatives that are highly dependent upon benefits in the distant future for their success seem hazardous. We simply do not know what humanity might do or create over the next fifty years, and therefore what buildings they might demand and how they might manage them.

1.6 Sustainable Decision-making

To decide whether a project or proposal is sustainable, you have to know the relevant definition of sustainability. There are a number of fundamental issues surrounding this. First, it is unlikely that all members of any project team will have the same understanding of sustainability, so conflict may arise. Second, sustainability involves the future, and the future cannot be known, so uncertainty becomes an important factor. Third, as a result, the components of sustainability (in the wider sense) can be given different levels of importance, and the extent to which they can be traded off can vary. Prudent members of decision-making teams should recognise that sustainability is a very complex concept, so can be defined in different ways, and that to achieve good project decisions some understanding of different positions and approaches is probably necessary. People (including clients and others involved in the project process) are fascinating decision-makers: on one hand capable of assessing and integrating complex and even vague information that can defy rational analysis, and yet also falling prey to a range of unfortunate biases. Recent research has helped to illuminate these matters. It is necessary to consider how people make decisions, and how multiple objectives can be met.

1: Notes

1   J. Joseph Sirgy (1982), 'Self-concept in consumer behavior: a critical review',
    Journal of Consumer Research, 9, pp. 287–300.

2   Ujjval Vyas (2009), *Sustainable Design and Construction: When Green Turns
    Red*, Chicago: American Bar Association, p. 4

3   World Commission on Environment and Development (WCED) (1987), p. 43

4   Ibid., p. 13

5   Ibid., p. 14

6   Ibid., p. 35

7   Sir Owen Williams (1927) 'Factories', *RIBA Journal*, 16 November, pp. 54–5

8   David Pearce and Edward B. Barbier (2000), *Blueprint for a Sustainable
    Economy*, London: Earthscan pp. 20–3

9   John Maynard Keynes, *The General Theory of Employment, Interest and Money*
    (1936), book III, chapter 8, part IV

10 WCED (1987), p. 24

11 David Pearce (2001), 'A bright green', *New Scientist*, 171(2309), 22 Sept, p. 50

You, the Clients and Other Decision-makers

2

2.1 People Aren't All the Same

2.1.1 Rational differences and biases

All architects must have had the experience of working for clients, regulatory agencies or consultants who want things to be done *this* way, when *that* way would be better – and who persist with this thinking despite their error being pointed out, whether politely or more firmly. This kind of difference of opinion can be uncomfortable and debilitating – it means that the architect cannot make decisions based on his/her own judgements, but instead has to imagine how someone with a different value system would make decisions. In this situation it's easy to make mistakes.

Differences can occur in relation to matters of taste, such as colour schemes or the style of door handles – which can be troublesome, but ultimately not very important. However, when it comes to decisions that relate to sustainability, for example the choice of a mechanical system, the client's value system and resultant priorities can be crucial. For an architect who is committed to sustainable design, it can be very frustrating to work for a client who seems over-cautious or uninterested. This is a common complaint of architects committed to delivering sustainability.

Should the architect try to change the client's mind? It's certainly valid to educate the client, to ensure that decisions are well informed and not based on out-of-date or erroneous assumptions. But there are powerful reasons why people make different decisions based on access to exactly the same data. To help reduce frustration, it's a good idea to understand why.

2.1.2 How do clients and architects (and others in the building industry) make decisions?

At a continuing education presentation on decision-making some years ago, one architect asserted that 'architects don't make decisions' – the notion being that they just did what they were told. It was a surprising assertion, but some members of the audience seemed to agree. It seems strange that years of formal education and experience would be required to create individuals who do not make decisions. Of course architects make decisions – even the most junior person makes decisions about which way to swing a door or where to end the tile and start the carpet. What this architect probably meant was that he rarely, if ever, undertook a formal, structured, decision-making process – collecting data, for instance, and putting it into a mathematical model. He meant that he did it some other way.

Given that architects, developers and managers really do make decisions, given the multitude of obstacles facing them when making decisions, how do they cope?

In the built environment, collecting accurate data and submitting it to detailed scrutiny and analysis is far from universal – in many parts of the industry it is still unusual. The basic tool of whole-life costing[1] is rarely used explicitly, and when it is used, it may not be to make decisions, but to fulfil government or client expectations. BREEAM, for example, rewards projects that undertake it – but there is no requirement that it be used in any meaningful way. For example, a partner in a substantial architectural firm commented that although they were frequently asked (in particular by government clients) whether they could do whole-life costing and they always replied in the affirmative, they had never been asked to actually undertake it. When asked what they would do if they were required to perform such an analysis the response was that they would probably buy a book on the subject. One large multinational engineering firm did routinely use whole-life costing, but commented that they used it to support decisions already made. They said that their clients appreciated having a quantified analysis showing why they should follow their consultant's recommendations. Perhaps the truth is that most of us would prefer some calculations, rather than just relying on an expert's gut feel.

Much decision-making (including building design) involves some sort of mysterious cognitive process, in which available information is sorted, factors prioritised, alternatives identified and refined, and iterations undertaken (traditionally with pencil in hand) until some direction emerges. Ideas may appear at the most unexpected times. The philosopher Ludwig Wittgenstein (1889–1951), who did more than the usual amount of thinking, suggested: 'A man's thinking goes on within his consciousness in a seclusion in comparison with which any physical seclusion is an exhibition to public view.'[2]

While in 1997 IBM's 'Deep Blue' beat chess world champion Garry Kasparov, and other computer-based systems can play other games, undertake house cleaning, drive cars and translate languages, even these tasks are simple compared with many that every architect, engineer and developer undertake. Questions such as 'What *exactly* am I trying to accomplish?', 'How should I balance a multitude of conflicting factors in this particular case?' and 'How do I decide when I have virtually no decent data?' can be faced every day. Game-playing computers know their objectives, have the rules of the game and can compute possibilities and probabilities because they also know the positions of the pieces and how they can move. The unfortunate designer or developer has to account for complex and changing situations, and the 'opponents' may be impossible to identify and follow a different set of rules. Unfortunately, there is no hiding; we all need to move onwards with projects, which means making and implementing decisions – hopefully sustainable ones.

2.2 Learning to Make Decisions

2.2.1 Informed intuition and expert judgement

If most property decisions are made using 'informed intuition', it is worth considering this concept further. Becoming a capable decision-maker can be a lengthy process. People move through stages of sophistication as they advance their capabilities in decision-making in their own fields.[3] There appears to be an integration of 'inquiry' and 'action' as one becomes increasingly competent in a field.[4] One model contains five stages that most experienced architects will recognise as career development stages:

1. **Novice:** This is the level at which individuals entering a profession learn and closely follow fundamental facts, rules and techniques. This does not necessarily just apply to younger people (such as recent graduates) – it is the way people tend to approach new subject areas.

2. **Advanced beginner:** During this stage, an individual begins to perceive fundamental patterns and relationships, and develops deeper understandings, which extend beyond the basic facts, rules and techniques.

3. **Competence:** At this stage, people start to recognise the true complexity of decision-making in their field and distinguish larger sets of cues. They will tend to rely less on the fundamental rules, and be better able to independently assess and take risks, and make better trade-offs.

4. **Proficiency:** As capabilities develop, informed intuition starts to dominate decision processes. A proficient decision-maker develops an ability to unconsciously assess situations, even within uncertain and evolving circumstances.

5. **Expert:** Ultimately explicit rules and absolute facts tend to disappear, and people will increasingly employ a comprehensive and multidimensional understanding of situations in the field in which they are expert. In some ways it is like learning to ride a bicycle or drive a car, only it usually takes years to develop – one does not have to think about how it works; it becomes somehow instinctive. At this stage, one is not necessarily a superstar – but becomes a true master of his/her field.

According to this model, designers and managers operating at the expert stage possess an inherent understanding of the issues and goals, and can approach problems in a comprehensive, integrated way, using whatever data can be reasonably obtained, combining it with their existing information base, precedents and some assessment of uncertainties. They can, using some unconscious cognitive process, create a set of feasible alternatives, and devise a reasonable course of action. Unfortunately, being an expert in one field does not necessarily make one an expert in another – something that can have unfortunate consequences.

2.2.2 Reasons why expert judgement develops (or does not develop)

The mysteries of the functioning and potential of the final-stage 'expert' decision-maker were demonstrated in a research project the authors undertook as part of a UK-DETR-funded project into the achievement of sustainability in the older building stock.[5] Two very experienced decision-makers (a property manager and an architect) were making curious decisions relative to the refurbishment of a relatively small Victorian hotel block used as a corporate conference centre. The yields from replacing the windows and ageing boilers were greater than those being accepted for the complete project. Basic analytical tools suggested undertaking the energy-saving work. The decision-makers were confident that they were following the best course of action by not replacing the windows and boilers. They could not explain why – they just felt they were right. Investigation by the authors eventually explained and fully justified their decisions (using real options; see Chapters 13 and 14), but it took some days to assemble the data and undertake the calculations required to replicate what they were doing instinctively. The two decision-makers had worked for the same large retail organisation for decades, and had repeatedly refurbished the company's buildings, so were in an excellent position to see the results of their previous projects. They had not collected and classified data, but possessed a great deal of experience, much of it presumably the result of trial and error. Unfortunately, perhaps inevitably, both retired within a couple of years and their expert capability was lost to their organisation and, ultimately, to society.

Unfortunately, some people may fail to develop into 'expert' decision-makers. The reasons for this include:

- **Not enough experience:** Gaining experience is an obvious prerequisite. 'Experience, however, is not a fixed state that can be created by following a recipe. You have to be lucky.'[6]

- **Organisational issues:** In pursuit of the maximisation of profits, not all organisations encourage learning. Young people in many organisations are just expected to follow orders.

- **Organisational compartmentalisation:** This is a particular risk in large organisations. Junior architects may spend years doing one task – perhaps door schedules or interior design – and may never be able to develop higher-level capabilities.

- **Outcomes are not observed:** The two corporate experts referenced were in an excellent position to observe the results of their decisions. Unfortunately, consultants are often engaged only for specific parts of projects, so do not see the longer-term implications of their decisions. While architects are often not in a good position to gain comprehensive insights into how their buildings are performing, in-depth post-occupancy evaluation, undertaken at different points during a building's life, will help to address this problem.

The challenge is to make people more capable decision-makers. It would be unfortunate if designers and managers became good decision-makers only in the last few years before retirement. Hence there is a need for explicit tools we can use earlier in our careers, to both make and understand decisions.

2.3 Bias in Decision-making: How We Cope

There are many experienced and capable designers and managers who may never need to use quantified tools. Some will be able to incorporate the latest technical and mathematical wrinkles into their decision-making without explicitly knowing about the techniques – and may even touch on methods not yet developed. Such individuals are capable of synthesising the innumerable factors that bear on a decision, developing alternatives and devising a good course of action. However, it can be difficult to identify those people who may possess this high-level capability – and many will be able to think of individuals who thought they were great decision-makers, even though supporting evidence was very weak. Moreover, in project meetings, the most capable decision-maker may not be the most convincing or the most powerful person in the room – hence the logic behind the previously referenced engineering firm making a decision and then verifying it mathematically.

Humans have also been frequently shown to be wilful and stumbling decision-makers, notably by Amos Tversky and Daniel Kahneman.[7] Our hunter-gatherer ancestors seem to have biased our decision-making in certain ways: for example, we tend to be wary of the unknown, and we also tend to run in packs – choosing

things that we know others have done before, or are accepted by our peers. This does not mean that such biases are necessarily completely inappropriate accompaniments to decision-making – but we do need to be aware of them, and recognise that they may not always be suited to the post-industrial context in which we live. An awareness of the ways in which our decisions can be biased can assist even the most capable gut-feel expert decision-maker to better calibrate his/her responses. A decision-maker should recognise his/her role in a decision-making process, and wonder why he/she might respond in a certain way.[8]

A complete list of the possible biases that may adversely affect decisions is dismayingly long. Many of these areas have been subject to considerable research, classification and sub-classification. A few might be considered.

2.3.1 Inertia (unwillingness to change)

Decision-makers have been shown to have a significant bias towards the status quo. There are several possible explanations for this.

First, it is often the easiest course of action – rather as one picks up a jar of jam or chooses a piece of door hardware that has been acceptable in the past. Little or no analysis is actually necessary. If the expected benefit does not outweigh the effort in making the decision, it is logical not to change – or perhaps, procrastinate.

Sometimes leaving the status quo means voluntarily taking personal responsibility for any change. Someone else made the old decision, but to change it risks two things – one might be blamed if things do not turn out right, and a failure may adversely affect one's own self-image.

Another explanation is found within subsistence economies – rather like those that have existed for most of human history. If one is close to starvation, any innovation, perhaps a new type of seed, has to be regarded with suspicion. The old methods may not perform very well, but they are less risky than the new – the consequences of adopting something new might be as catastrophic as crop failure and your whole family starving to death.

The 'endowment effect' can also help to explain inertia. Experiments have demonstrated that when someone already owns something, they assign more value to it than when they do not.[9] This is a complex area in itself, and encompasses ideas as well as physical goods. Architects all know what it is to fall in love with a design, and resist abandoning or changing it – even if it is just a rough concept sketch.

2.3.2 Anchoring and adjustment

The concept of 'anchoring' has been well documented in many experiments. Anchoring encompasses biases created by past events, whether distant or recent. A typical experiment asks whether some factor, perhaps the population of a country or the price of a good, is more or less than some number. The second question asks for an estimate of that factor. Typically the resultant estimate is biased according to the number offered in the first question. Logically, some randomly selected number should not reasonably change the estimate made, but in practice it usually does.

One area in which anchoring can be used exploitatively is in the marketing of art. What is a painting worth? The answer is: it depends. In the case of art, where the fundamentals determining value are so elusive, anchoring can be a very effective selling tool. A recent painting by an established but not well-known artist might be offered in a gallery for a couple of thousand pounds. Through astute negotiation, someone buys it for two-thirds of that price, and thinks he has a great deal. Some years later, the purchaser's niece is selling his household effects and the painting is unknowingly offered for fifty pounds, and someone negotiates her down to forty pounds – a good deal again. It is the same painting, only the context (perhaps a boot sale) and the starting price establish the value. The opening price worked to anchor the potential purchaser. It unduly influenced what someone is willing to pay.

In the building industry this can be one reason why capital budgets are so often inadequate. A project inevitably starts with a sketch capital budget – perhaps somewhat on the low side to support the feasibility and validity of the initiative. As things proceed, more details become known, so it can become more accurate; however, decision-makers will always be haunted by the memory of earlier budgets. Funding may never be brought up to a realistic level, resulting in anguish as the final costs roll in.[10]

Anchoring can also deal with stereotypes about people or types of businesses or products. Essentially, an anchoring bias gives too much weight to past information – it may be relevant, but it might also obscure other data which is more recent and of higher quality.

2.3.3 Sunk costs

The problem of sunk costs appears in a number of areas of decision-making. Sunk costs are those that have already occurred – resources that have already been expended. In a decision-making situation it tends to mean that one possible direction has received some funding already, and if that alternative is not selected the money cannot be recovered. Sunk costs should in theory be irrelevant to decision-making, but most people and organisations find them hard to ignore – even though we have a popular term that might refer to such cases: 'throwing good money after bad'. Logically, only future costs and outcomes should matter; whether or not money was spent in the past should be immaterial. In some situations one possible alternative may have been advantaged by past expenditures, but the important thing is to look into the future when evaluating project alternatives.

Sunk costs can influence a decision inappropriately; the reluctance to abandon sunk costs can even seem like rational behaviour from some viewpoints. This is because writing off sunk costs may highlight a wrong decision. Not changing course may mean that no one ever notices that there were better ways to do things. It can be tempting to defend a design or a product even after it has become apparent that another approach would be better. For consultants, writing off sunk costs may mean writing off design time, with the client receiving the benefit of the redesign but the consultant paying for it. Of course, the matter of sunk costs is interwoven with inertia, and with excessive optimism – a design solution which is already in progress has considerable advantage over other alternatives, even though it may not have a high relative ranking.

2.3.4 Confirmation bias (selective search for evidence)

In a complex world, full of ever-increasing volumes of facts, statistics and 'evidence', there is a serious risk of searching for, finding and/or interpreting information that confirms one's preconceptions, and of avoiding information that might contradict them. Experimental evidence has shown that decision-makers will tend to accept uncritically and assign more consequence to information that confirms their existing hypotheses, and give less weight to evidence that goes against them.[11] The electronic media we watch and the magazines we read tend to align with and support our pre-existing beliefs. People who have a conservative orientation rarely read socialist newspapers – there is a natural tendency to avoid what we dislike.

2.3.5 Peer influence

As social creatures, humans often have their decisions influenced by social context. The concept of 'groupthink' is well established, having emerged in the early 1970s – the propensity of people to fall into harmonious agreement and to suppress non-conformist creativity and enquiring questions. Indeed, in many organisations expressing dissent can be hazardous to one's career.

We do not have to be members of groups sitting around a table looking at a particular project to fall prey to peer influence. Fascinating research undertaken by Salganik, Dodds and Watts (2006)[12] demonstrated experimentally, using downloadable musical tunes, that we can be strongly influenced merely by what we know about how other people are acting – even those we don't know. Such experiments help to explain various phenomena – the emergence of curious fashions, the importance of early market dominance, the persistence of 'blockbusters' – but most importantly how incredibly influential peer influence can be. Simply knowing how others out there are making their decisions can be enough.

2.3.6 Biases in frequency judgements

We are frequently attempting to assess the likelihood of events as part of a decision-making process. Studies have demonstrated that people tend to underestimate very high frequencies and overestimate very low frequencies. This has been found to be true with respect to people estimating the frequencies of diseases – substantially overestimating the possibility of dying from botulism poisoning, for example, while underestimating the likelihood of dying of a stroke.[13] Humans tend to overestimate the risk posed by rail accidents (very low), and underestimate that of car accidents (much higher). This bias can to some extent be explained. Unusual events, such as tornadoes, earthquakes or nuclear power station accidents come to your attention and remain in your memory, simply because of their rarity. When making a decision, those unusual and recent events may distort thinking and affect probability judgements.

2.3.7 Unconscious over-claiming/over-confidence

We tend to unconsciously overestimate our skills and capabilities relative to those of other people. This is probably logical, if for no other reason than we are more aware of them, and how we got them. Again, research shows that we tend to believe that we are more capable of making estimates and forecasts than we actually are and that we have more control over outcomes than we really do. In particular, we often tend to ignore the role of uncertainty, and of luck. Many people in the property industry will underestimate the importance of uncertainty – not recognising that ongoing fluctuations of key variables will affect a project's success. This means they may not fully plan for how to deal with possible planning difficulties, poor soil conditions, unexpected costs or changing markets. Built projects, in particular large ones, may take years to bring to market, and by then market prices or rents may have fallen. Usually the expected or most likely outcome is not the only possibility. The failure of major property development companies is far from unknown.

2.3.8 Optimism/pessimism and our fascination with doomsdays

One factor that is very important when considering how we deal with the future can be expressed in terms of individual levels of optimism and pessimism. This is a significant element of human behaviour and is a field of study in both psychology and business. Each of us holds beliefs about how things will turn out.

Much evidence suggests that certain aspects of the human brain are biased towards optimism. This can affect how people assess risk – rejecting the possibility of bad things happening, as well as overestimating benefits and underestimating costs. Optimistic people have been shown to be more successful, because they have higher expectations and a greater propensity for action – leading to them putting more effort into projects. It might be suggested that the observation of the more entrepreneurial property developers would support these propositions in both positive and negative ways. Management research has also found that entrepreneurs tend to be more optimistic than managers who work within organisations.[14]

The development and construction industry seems to be the natural habitat of optimists, and it's likely that everyone in the industry has encountered them – or is one themselves. Sometimes these individuals have experienced numerous project disasters, even financial failure, yet are still willing to take on yet another high-risk venture. The public tends to notice those who are either the most capable or just lucky – just as casinos and lotteries celebrate the winners, not the losers. Confident, optimistic people tend not to talk about their failures – they are too busy plotting their next project.

However, there is a downside: over-confidence, which has been found to be associated with poor decisions. This may be as a result of risks not being identified or assigned enough significance, not enough effort being put into the acquisition of information, the suppression of dissenting opinion, overestimation of personal capabilities and/or not using available analytical tools.

Curiously, we are also fascinated by the concept of doomsdays. Much sustainability literature is framed in terms of doomsday scenarios. Reading material from the past can be revealing, partly because we are now well past many 'best before' dates, and not only does humanity survive in a more-or-less civilised state, but by many measures it is better off than when the predictions were made. For instance, there have been very significantly declining levels of global malnutrition over the past decades.[15] Apocalyptic world views are often framed in sales terms that are intended to lead the reader to believe that events such as mass starvation and a collapse of civilisation are inevitable unless we follow some particular strategy.

The point is that the decision-maker, as part of the human species, has to recognise his/her own inherent biases, especially when someone is trying to sell him/her some product or concept. As with other human biases, the aim is to achieve some level of balance.

2.4 How Should We Evaluate Alternatives?

Every participant in the building world can probably provide an example of when one or more of these (or other) biases have affected a decision, and anyone with any significant experience will probably be able to ruefully recall when he/she personally fell victim to this.[16] Keeping them under control is not the easiest task, because they are a part of human nature. Even the best gut-feel decision-maker can use some calibration from time to time, and there will always be the need to convince others of your judgement – explaining to a client committee of MBAs that you are 'following your instinct' may not always be enough.

The most widespread discussion of biases revolves around the boom-and-bust cycles of the stock markets and other similar markets. After the fact, descriptions of the processes observed often include words such as 'euphoria', 'mania', 'distress' and 'revulsion', referring to what should be serious, rational investment decisions involving billions of pounds/euros/dollars.[17] Architects do not always see themselves as making investment decisions, yet the process of design does constitute an allocation of significant economic, social and environmental resources. Without suitable tools, it is not easy to have insights into what constitutes an appropriate decision: some people are very capable salespeople, and one would hope that not all decisions are made on the basis of who is the best at swaying an audience, using his/her ability to convince in visual or verbal presentations, or simply by pheromones or the force of personality.

The ultimate problem is that we may not evaluate alternatives rationally. If we are not careful, we might direct substantial resources to address matters that offer small, perhaps almost insignificant, returns and miss better opportunities.

2.5 Where to From Here?

Researchers have told us that all humans – including those in the property world – can be wonderful decision-making machines, but also carry an apparently inexhaustible collection of biases. We know that most architects' decisions do not involve the explicit collection of data and the use of mathematical techniques, but result instead from fuzzy cognitive processes where vast collections of inexact and incomplete data are reviewed, anecdotes considered, and alternatives developed, twisted and turned – all of which lead to eventual resolution (sometimes, as in the case of Archimedes, in the bath). Explicit analysis often does not occur, sometimes for good reason. The practitioner, recognising this, needs to feed his/her inner decision-making processes with additional experiences, insights, and analytical and conceptual techniques to: 1) assist in dealing with his/her inherent biases, 2) be able to understand why other people on a project team may come to different conclusions, 3) improve 'unconscious competences', 4) help to understand when a more extensive quantified analysis is necessary and 5) be able to perform quantified analysis – including the usual tool, whole-life costing.

2: Notes

1 Whole-life costing is used to express the application of discounted cash flow analyses and associated techniques in the built environment. Other terms commonly used include life-cycle costing, discounted cash flow analysis, and cost–benefit analysis.

2 Ludwig Wittgenstein, *Philosophical Investigations* (1953), p. II, p. 189

3 W.R. Torbert (1987), *Managing the Corporate Dream: Restructuring for long-term success*, Homewood, Ill: Dow-Jones Irwin, and Robert E. Quinn (1988), *Beyond Rational Management*, Oxford: Jossey-Bass

4 Quinn (1988), p. 7

5 Department of the Environment, Transport and the Regions

6 Quinn (1988), p. 23

7 Kahneman, a psychologist (not an economist), won the 2002 Nobel Prize in Economics

8 An extensive discussion of human biases can be found in Daniel Kahneman, Paul Slovic and Amos Tversky (eds) (1985), *Judgement under Uncertainty: Heuristics and Biases*, Cambridge: Cambridge University Press

9 Daniel Kahneman and Shane Frederick (1990), 'Experimental tests of the endowment effect and the coase theorem', *Journal of Political Economy*, 98, pp. 1325–48.

10 Dan Lovallo and Daniel Kahneman (2011), 'Delusions of success: how optimism undermines executives' decisions', pp. 29–49, in *Harvard Business Review: Making Smart Decisions*, Boston: Harvard Business Review Press.

11 Charles G. Lord et al. (1979), 'Biased assimilation and attitude polarization: the effects of prior theories on subsequently considered evidence', *Journal of Personality and Social Psychology*, 37(11), pp. 2098–109.

12 A summary appeared as: Richard Webb, 'Online shopping and the Harry Potter effect' (2008), *New Scientist*, 22 December 2008, issue 2, p. 687

13 S. Lichtenstein et al. (1978), 'Judged frequency of lethal events', *Journal of Experimental Psychology: Human Learning and Memory*, 4(565), from Baron p. 138.

14 Rose Trevelyan (2008), 'Optimism, overconfidence and entrepreneurial activity', *Management Decision*, 46(7), pp. 986–1001.

15 Mercedes de Onis and Monika Blossner (2003), 'The World Health Organization Global Database on child growth and malnutrition: methodology and applications', *International Journal of Epidemiology*, 32, pp. 518–26.

16 Daniel Kahneman himself commented in Tim Adams (2012), 'This much I know: Daniel Kahneman', *Guardian*, 8 April, 2012, http://www.guardian.co.uk/science/2012/jul/08/this-much-i-know-daniel-kahneman, accessed 11 September, 2012: 'My main work has concerned judgement and decision-making. But I never felt I was studying the stupidity of mankind in the third person. I always felt I was studying my own mistakes.'

17 Niall Ferguson (2009), *The Ascent of Money: A Financial History of the World*, New York: Penguin, p. 123

Sustainable Buildings

3

3.1 What Makes a Building Sustainable?

Suppose that we want to design a sustainable building. Architects often focus on what materials or technologies they can specify, or what features they can include, as ways to enhance sustainability. But sustainability isn't assured on the day a building is completed; it is only revealed through long-term performance. More thought is required.

A building is sustainable over its lifetime if it continues to make a positive contribution to physical capital, with the activities taking place in it adding to human well-being, and without it being an excessive drain on natural capital.

In contrast, premature obsolescence will undermine the sustainability of a building. This could happen in a number of ways. First, if it is technically sound, but no longer serves a useful purpose that contributes to human well-being; second, if it continues to serve a useful purpose that contributes to human well-being, but it is technically worn out or has unaffordable running costs (financial or environmental); or third, if it is no longer esteemed, so is allowed to deteriorate.

It should not be surprising if this sounds vaguely familiar – it echoes an ancient interpretation of good architecture, going back centuries, but expressed in English in Sir Henry Wotton's *The Elements of Architecture* (1624) as 'commodity, firmness and delight'. In a modern sustainability sense, ways of satisfying these three factors over very long time frames have to be considered.

An example of the first kind of premature obsolescence would be a boarded-up shopping centre, solidly built but empty; an example of the second kind would be a primary school with draughty, uninsulated classrooms with leaking roofs. Both of these buildings probably worked well, were safe and looked good on the day they were finished, but experience over time revealed their design to be unsustainable.

Designers do not have the benefit of being able to see into the future to predict how their designs will perform – so how can they make their designs sustainable?

3.2 Underinvestment, Overinvestment and Efficient Investment

When architects specify their designs they take account of how a building will perform after completion. Architects are faced with many specification choices, as succinctly expressed by the French professor of architecture Jean-Nicolas-Louis Durand in 1819: 'Buildings are made of a wide variety of materials, but they fall into two classes: 1st, those that are more durable but more expensive; 2nd, those that are less durable but cheaper. One uses the first class in more important buildings and the second class in less important buildings.'[1] When today's architect or developer considers alternatives this is a very common question – more durable materials and techniques typically cost more than short-life alternatives.

Designers should choose the right materials for the right buildings – this leads to efficient investment and avoids the following pitfalls:

- **Underinvestment:** This occurs when a low level of specification is used for a long-lasting building, with the result that the use of the building is hampered or there is a heavy burden of maintenance or component replacement that could have been avoided with a higher initial specification.

- **Overinvestment:** This occurs when a high level of specification is used in the expectation of benefits through the building's life – but these benefits do not actually arise because the use was short-lived or less demanding than expected, and a lower initial specification could have performed adequately.

Neither underinvestment nor overinvestment make the optimum use of resources, so both fail with respect to sustainability. With underinvestment, the current generation is too stingy and as a result imposes a burden on future generations. When today's new buildings are criticised for being unsustainable it is usually because there is believed to be underinvestment. Advocates of sustainability generally prefer to see more invested in insulation, renewable energy systems, long-lasting and recyclable materials and so on. Yet it is not that simple. High upfront investment increases the risk of overinvestment, which can be just as unsustainable as underinvestment. The resources used for investment by today's generation are not available to future generations; if the benefits are disappointing they can rarely be recovered.

When aiming for efficient investment, it is helpful to be alert to the risk of both underinvestment and overinvestment. The idea of overinvestment may be less familiar so it may require extra attention when attempting to develop a sustainable design proposal.

3.3 Why Buildings are Different

Architects who genuinely desire to create a more sustainable built environment have a serious mission. Buildings really are different from things like cars, clothes and cans of juice. One implication is that much of the work done on sustainability in other contexts does not immediately apply to buildings and land. Decisions about the sustainability of the built environment need to take into account the factors that make buildings and land different from most other goods and assets.

- **They have multiple attributes.** Buildings and real estate are 'bundles of attributes'. A typical house offers a range of benefits, including shelter, status, personal expression, a financial asset, access to schools of choice and proximity to employment. In contrast, a pair of shoes, for example, might offer two main attributes – comfort and a fashion statement (or perhaps only one of those).

- **They are fixed in space.** Most assets can be moved to locations where they perform best, but not buildings. Suppose an entrepreneur were to build a ferry terminal and buy a boat to take visitors to an island. If a bridge is built, the ferry can be moved to a different route, but the terminal building may become useless. The value of buildings is entwined with their location. Being fixed in space, they are targets for regulation and taxation because they cannot move to a more lenient jurisdiction.

- **Every building is unique.** Buildings are all individual. Inevitably, each one occupies a different site, but individuality of use must also be taken into account. In the case of a semi-detached pair of cottages in the Cotswolds, one may be lived in all year and the other may act as a second home that is only occupied in the summer. For each house the appropriate insulation and heating strategy would be very different. For the occupied house, external insulation might work well – but not in the Cotswolds where conservation of heritage trumps conservation of energy. Although there are some general principles of sustainability, they must be applied on a case-by-case basis.

- **They have a very long life cycle.** Almost everything associated with buildings takes longer than for other products. Buildings take protracted periods of time to create and they can also last for protracted periods of time – at the current rate of construction and demolition in the UK, it could take 1,500 years to replace today's housing stock. Most buildings in use today were originally designed for a very different world, and many have experienced significant change of use. A violin, no matter how old, will probably see service only as a violin, and juice cans are simply disposed of when empty, but a house can become an office, an office building can become a block of flats, and a power station can become an art gallery. The values and motives that applied at the time of an older building's creation are of little interest to today's decision-makers, who have their own values and motives, and the same is likely to happen with the designs we make today in ways we cannot even guess at. Almost anything involving buildings takes time; typically years pass between the initial idea and the completion of an office building.

They are universal. Practically all human activities require buildings, so almost all human beings are building users. Everyone lives somewhere, is influenced by the built environment and often has something to say about how it's managed. People can be so willing to express their opinions that in one of the authors' neighbourhood surveys, more questionnaire forms were returned than distributed. Unlike an architect, the designer of the next-generation car does not worry about NIMBY (Not In My Back Yard) objections.

There is open-ended demand for them. The demand for buildings often seems to go far beyond rational need. Individuals and corporations invest heavily in buildings for reasons that are best interpreted by the psychologists. For most goods, the 'law of diminishing marginal utility' means that demand tends to taper off as more of a good or service is received. The gain in benefit felt from an extra £1,000 is much less for a person with lots of money than for a poor person. However, the demand for buildings, housing in particular, can be insatiable: people's appetite for more or bigger buildings seems to grow without limit corresponding with the amount they can afford (or think they can afford). Even after some fundamental demand for basic shelter has been met, other, less tangible aspects, such as the fulfilment of the need for personal status, can remain unmet. This curious factor has sustainability implications. A wealthy and sustainability-minded client may want a zero-carbon second home, but surely it would be more sustainable not to have a second (or third) home at all. How far does the role of the architect extend? One might incline to the view that the architect's job is to satisfy societal demands as skilfully as possible, not to be responsible for whether the demands are truly rational in the great scheme of things.

They are subject to cultural relativity. The built environment both reflects and creates much of our culture and the way we choose to live. Why don't people wear more clothes in the winter, instead of heating their homes to summer temperatures? Why do people in the tropics build glass buildings with enormous air-conditioning loads? Why do people in earthquake regions rebuild their collapsed buildings using the same, vulnerable technology? Architects can make suggestions for how to do things better (as seen from their point of view) but the take-up will be constrained by their clients' perceptions of what is desirable and possible.

■ **There is a lack of comparable data.** For many reasons it is difficult to get good, useful data on how buildings perform. There are many variables and little transparency on topics such as occupancy patterns, fuel consumption, refurbishment, management style, rents, sale prices and so on. The uniqueness of each piece of property means that exact comparables are unusual. Only broad-brush generalisations are possible (and these can be useful, as far as they go). For the architect, every case is a special case.

All of these factors may lead to many difficulties in decision-making, but it is the world in which architects and developers have to operate. Ignoring these constraints is more likely to impair than improve the decision-making process.

3.4 Buildings and Uncertainty

Buildings exist in a context of uncertainty, but not all buildings (or building components) face the same uncertainties. If the future could be known, it would be easy to make decisions about how much to invest to achieve a project's objectives, to obtain maximum value over its lifetime.

Uncertainty makes the achievement of balance between underinvestment and overinvestment difficult. This is where explicit tools come into their own. The use of discount rates is a primary tool, together with other techniques such as real options approaches, all of which coalesce in whole-life costing. Decisions relating to buildings and building elements that face high levels of uncertainty should reflect high discount rates, which imply short-term planning. While this sounds inappropriate ('short-termism' is popularly seen to be a reprehensible strategy), in contexts of high uncertainty it can be appropriate. If the decision-maker cannot count on future returns materialising, he/she should not pay as much to get them as if they were certain to appear. Why use expensive flooring in a situation where it might be ripped up in a few years?

The use of discounting procedures is often criticised, especially concerning matters sustainable, but that is often because they are treated as 'black boxes' that cannot be opened up. However, careful and informed choice of discount rates can lead to numbers that encapsulate the characteristics of a project and its environment.

Some of the primary factors of uncertainty affecting projects might be considered, as follows.

3.4.1 Long life or short life?
Building and component life are key factors in making sustainable decisions. Some buildings and building elements tend to have long lives, while others do not. This can be affected by a number of factors.

3.4.2 Stability of use

Houses are one type of long-lived building, and in many developed countries now experience a very low ongoing rate of demolition. This is different from a few decades ago, when poorly constructed housing built in the early 19th century was being replaced. The appearance of building regulations and planning regimes has tended to increase the life expectancy of houses in the UK. House designers would be wise to take a long-term view, even though individual owners may have a short-term interest.

Conversely, restaurants often have a very short life – many fail within one year of opening. That suggests planning for a short life. If the restaurant becomes one of the relatively few that do last for decades upgrading is possible.

Of course, it is widely recognised that different elements in a building may face different levels of stability of use. While the structure of a small shop might be expected to last many decades, the interior fittings are not likely to. Even if the use remains constant, attempts to remain in fashion may require periodic replacement.

3.4.3 Stability of regulations

Buildings and their use exist in a context of regulation. Different uses are subject to different uncertainties with respect to regulation. For example, rising expectations about healthcare and long-term care for the elderly mean that facilities built for those purposes may be forced to conform to new government-created requirements, no matter how well the building is fulfilling its original mandate.

3.4.4 Stability of technology

Some buildings relate to technologies that may be rapidly changing. Warehouses, for example, have evolved to incorporate automated storage systems.

3.4.5 Link between construction and operations

How an organisation is ordered and controlled can have a significant impact on how its people make building-related decisions. If it has a divisional structure, it is likely that energy and other building concerns are seen as minor matters when compared with staff costs, the supply of raw materials or marketing. In a functional arrangement, an estate management department might be responsible for buildings and another for staffing. In such a structure, the people making the decisions about buildings will see energy and maintenance issues as important budget items, so pay more attention to them. In addition, building management regimes do vary widely; the complex mechanical system that works well for an organisation with a capable staff may not work well if staff are poorly trained or transient.

3.5 Significant Points to Remember

3.5.1 Sunk costs

'Sunk costs' are costs that have already been met and cannot be recovered, and are always a major hazard, to be treated with great caution. In most cases they should be irrelevant to the decision about whether to continue with or abandon the project.

Should a partially complete design be continued? The decision is whether or not to commit more resources. There aren't any decisions to be made about the resources that have already been expended. It may take a conscious effort, but the focus should be entirely on future benefits, ignoring past costs. An abandoned initiative represents a loss, and potentially an embarrassment, but things could be even worse if additional resources are committed to a sub-optimal scheme.

3.5.2 Opportunity costs

Another powerful idea that at first seems alien to our emotions and gut feel is the concept of 'opportunity costs'. The basic principle is that you can never establish the true value of a proposal in isolation, without a frame of reference established by other possibilities. To make a good decision it is essential to assess alternative courses of action, because a consequence of selecting one course of action is that alternatives are excluded. Limited resources might mean, for example, that the budget for a hotel upgrade could allow for the refurbishment of the bedrooms or the conference suite but not both. In other cases, regardless of resources, choices are mutually exclusive – it is difficult to clad a wall (or at least an entire wall) with both marble and granite, no matter how much money you have, and the designer has to choose between building a roof out of timber, steel or concrete. The next best alternative not selected is the source of the opportunity costs. The designer should try to determine what choice or choice set will create the highest increase in value for the building.

Some sets of decisions are relatively easy, because they can be brought to some equivalence. In the case of others there is no common yardstick of value. One might consider a few hectares of open land in an urban area. It is possible to calculate the development value of such a property by establishing how much it would command in the marketplace. If the land is used as a park it will have amenity benefits for the people who use it – peace, tranquillity and a place to walk the dog. These benefits are undoubtedly of value but are difficult to quantify or express monetarily. Moreover, some things have value to people who have never encountered them, rather as Central Park in New York might be esteemed by people who have never been to New York. Each use of the land has opportunity costs due to the excluded alternatives. It is impossible to calculate the best use of the land without knowing society's values; often the best way of finding out what society's values are is to observe whether the land is developed or landscaped.

This question appears in many places. For example, given limited resources, what environmental strategy makes the most sense? Should the designer worry about energy consumption or waste production? Resources expended on one strategy cannot be used on another. A recent case was that of a proposed small-scale hydroelectric generating plant in a watercourse that had been extensively rebuilt by the Victorians and subsequent generations. The project had been held up for years because eels had been found in the river. The costs of protecting the eels made the project uneconomic. Hence, the decision process included having to determine whether to protect eels or have a source of renewable CO_2-free electricity. In such cases, available analytical approaches are often bypassed and debate becomes politicised, and can be 'won' by the side with the greatest ability to influence the bureaucrats, the electorate and the politicians.

Green-oriented regulations are great at rearranging opportunity costs. They require everyone to do something that might be good as a general principle, but at the cost of excluding alternatives that may be better in particular cases. For example, if conventional tungsten filament light bulbs are banned, you might have to put a compact fluorescent lamp in a cupboard where the light is only turned on for a few minutes a month. The energy saved may never repay the embodied emissions that went into the compact fluorescent lamp. Market-based schemes, if they work well, allow investors to choose between many alternatives and select one with benefits that exceed the opportunity costs.

Opportunity costs are easy to overlook. A decision-maker may choose a good strategy without realising that it incurs opportunity costs that outweigh the benefits, by excluding an even better strategy. Being alert to opportunity costs is the equivalent of always asking, 'Is there a better alternative?'

The reality is that most opportunity costs are difficult to assess and even more difficult to quantify. Sometimes opportunity costs are identifiable with hindsight – if only I had done that …' – but at the time the decision is being made the significance of many opportunities is not knowable. The unfortunate reality is that decisions must be made, data is poor, uncertainty is rampant – and not every alternative can be recognised and fully assessed. Perhaps the best thing is to work to identify alternative possibilities, assess them to the extent possible, try to ensure there is some flexibility whenever possible, make the decision … and move on.

3.5.3 System boundaries

When evaluating a strategy, how far should you go in looking for alternatives and assessing the opportunity costs they impose? That is to say, what is the boundary of the system within which alternatives can be sought?

Consider a pair of semi-detached houses – one occupied by an eco-enthusiast who has made a lot of energy-saving improvements, and the other by a procrastinator who has made none. Having done the simple, cost-effective things, the enthusiast still has funds and enthusiasm, so he plans to embark on an expensive project yielding negligible financial or energy returns. But his neighbour hasn't yet done the most basic, cost-effective things, so the eco-enthusiast would achieve more eco-benefit for his money if he paid for basic upgrades in the procrastinator's house. Is this a realistic alternative? Does the system boundary for the eco-enthusiast include or exclude upgrades to his neighbour's house? As with Sir Owen Williams' line of questioning about factories, one might wonder about the boundaries of the system. How far does one look?

For example, regulations might require a certain percentage of the energy consumed in new buildings to come from site-based renewables, in order to reduce CO_2 emissions. This can lead to high, and probably inappropriate, expenditure on an inefficient energy-generating plant, because the funds could be used in other ways to give much greater CO_2 savings, though this would require a change to the system boundary. Another system boundary example might be installing photovoltaic panels in a desert region. They will produce a lot of electricity, but they will get dusty and will have to be washed frequently. If the water supply comes from energy-intensive desalination plant, the energy to create the fresh water may undermine the benefits of the energy produced.

3: Notes

1   Jean-Nicolas-Louis Durand (2000), *Précis of the Lectures on Architecture*, Los Angeles: Getty Foundation (first published in Paris, 1821), p. 189

Looking for Ideas About 'Needs' – Some Thoughts from the Economists

4

4.1 Economic Concepts and the Sustainable Decision-maker

Any designer or manager has to try to assess the various messages that are put before him/her offering sustainable or green products, systems and ideas. A major issue is that while some of these might be very appropriate in certain contexts, they may be completely inappropriate in others. Fortunately, a number of concepts, largely derived from economics, can assist in independent assessment of not just what to choose, but when to choose it, how much to choose and how to mix elements together.

One of the difficulties in understanding sustainability relative to buildings is that buildings are complex assemblies of interdependent systems. Numerous components, attributes and external conditions work together, influence each other and, in the context of multiple sources of uncertainty, collectively lead to what may be unexpected consequences. Designers and managers need to consider the many forces that may influence outcomes over time. The final product has to attempt to achieve a meaningful integration of a stunning variety of factors within an ever-changing setting.

Economics as a discipline only exists because of scarcities – things that are in limited supply – and that includes almost everything. This is why, in the search for better ways of making sustainable decisions, one of the most promising areas for the sustainable decision-maker to explore and exploit is economics. Economists attempt to understand, in a structured way, the production, distribution and use of natural resources, goods, services, labour and, inevitably, money. As with the sustainability ethos, economics concentrates on understanding the factors that contribute to human well-being. Although Adam Smith (1723–90) is widely regarded as the father of modern economic studies, the field has its roots in antiquity.

Presenting a comprehensive background in economics is beyond the scope of this book; however, some key concepts should be of particular interest to the architect making sustainable decisions, and it is worth looking at them a little more closely.

4.2 What Gives Things Value?

Any exploration of sustainability has to enquire about how to maximise well-being, individually and collectively, now and in the future, so it is only logical to consider how well-being is formed. The early economists had the same concern, but expressed it in terms of value and attempted to understand how economic systems allocate resources. Why did goods and services have the worth that they did? Why would people pay different amounts for them? Adam Smith offered useful opinions on the matter, but his contemporary, the poet Robert Burns (1759–96) was more concise.

> Written on a pane of glass in the inn at Moffat.
> *Ask why God made the gem so small,*
> *An' why so huge the granite?*
> *Because God meant mankind should set*
> *That higher value on it.*

Unfortunately the wisdom of Burns has often been ignored. Some early economists saw value in a rational, physical-scientific way, much as a piece of industrial equipment might be valued in terms of its ability to produce something. Of course, people working in the built environment will immediately recognise the fallacy of this method of valuation – it misses much of what architects (and marketing managers) strive to create. To the designer or marketing manager, it is not difficult to understand that much value results from a subjective response on the part of consumers – the outcome of personal value systems. For example, High Victorian buildings, although esteemed by their creators, were seen as blots on the landscape through most of the twentieth century – assets of little value, and often something that could not be disposed of quickly enough. More recently, they have become things to be cherished and protected. Their value to society and in the marketplace shifted dramatically without the asset itself changing much; instead public preferences changed.

Command economies, such as that of the Soviet Union, set prices according to policy – one common strategy has been to keep oil prices low. While this assists industries dependent upon oil inputs and is popular with motorists, it does not encourage efficient usage. Low fuel prices undermine energy-saving investment. Complex actions of the marketplace result from the actions of millions of suppliers, consumers and intermediaries, and ultimately determine the relative values of different goods and services. The free-market economist Friedrich Hayek saw market allocation mechanisms as integrating more information than bureaucrats and politicians could possibly deal with, suggesting that decisions made as policy instruments would inevitably fall short of those made in a free marketplace.

Adding the concept of sustainability makes resource allocation even more difficult, because it forces people to evaluate alternatives that can be difficult to express using any shared measure. A caution lies here for the sustainability-focused decision-maker: remember that people with agendas may set values and equivalences to further their own aspirations. As with the Soviets setting prices, this may lead to sub-optimal outcomes. While all of this can become quite complex, some basic concepts should be of interest to the decision-maker attempting to enhance the sustainability of his/her buildings.

4.2.1 What people want: how we reveal it

The question 'what is the value of a good or service?' remains relevant when one moves into the field of sustainability. There are a number of theories, but economists generally accept that how people make choices in the marketplace (also known as 'revealed preference') offers useful insights into how people define 'needs'. Preferences can be observed: what bundles of goods and services people select, and how they treat them relative to each other.

An important revelation from observing human behaviour is that assigned value is not a constant, but is subject to many influences, including number and characteristics of consumers, availability and price of substitute goods and services, changes in the availability and price of associated products (i.e. heating systems require fuel), consumer incomes (rising incomes cause us not just to consume more, but differently), changes in preferences (fashions and trends) and future expectations with respect to price and availability. Demand factors interact with availability (supply) to form some structure of preference, often expressed in the form of price. To a reasonably educated and insightful person, all of this should be self-evident, and seen to be widely applicable to both tangible and intangible goods and services, including those that are associated with 'green' initiatives, such as pleasant urban environments, rural vistas and clean air.

4.2.2 The affluent society

The benefits received from goods and services are influenced by what goods and services have already been acquired. In a sustainability context, this relates back to the problems in expressing well-being only in terms of per capita GDP. In developing cultures, people strive for material goods – basic shelter, transport, food and furniture. Gradually, as these needs start to become fulfilled, the marginal benefit of acquiring more of those goods and services falls, and other priorities become relatively more important. Needs and desires that appear and increase with affluence are often for things that tend to be less material. This was explained by Abraham Maslow (1908–70), whose higher levels of aspiration include love, belonging, creativity, self-esteem, respect of others, friendship and family – things less directly linked to money than such basics as food and shelter.

4.3 Diminishing Returns and Marginal Utilities

The benefit resulting from any asset varies depending upon context. One of the most important elements of context is how much is already possessed. The amount of benefit derived from one additional unit of a good or service is termed the 'marginal utility' and follows what is often termed the 'law of diminishing marginal returns'. This function relates both to how much of the good or service has already been used or is available, and also – usually – to the amounts of other goods used or available. Eating one chocolate bar is pleasurable for most people, but eating the fifteenth is rarely so. Nor is it pleasurable if one has previously eaten an entire pie. Usually the additional benefit gained from an additional unit of a good or service is inversely related to the number of units already available or the availability of near substitutes, for example a more relaxed office dress code relative to mechanical cooling, or better heating and warmer clothing. This concept underlies much about what we understand about how well-being is expressed, and is important in any consideration of how to design for sustainability.

As a result, benefits received tend not to form a linear relationship with resources expended to get them. This concept helps to explain some observed behaviour. One widely used example concerns lotteries, which are often criticised as being a 'tax on the poor'. Research has shown that manual workers and the unemployed are significantly more likely to play lotteries than the affluent. One reason for this is that the marginal utility of money (acting as a proxy for all sorts of goods and services) varies from person to person. If a poor family won £100,000 it could make a considerable difference to their lifestyle – they might take a long-postponed holiday or buy a house. To a more affluent family, the money would still be nice, but would not cause a radical change – they probably already take vacations and own a house. Although the money might allow them to buy a bigger house, the difference between owning a house or not is much greater than the difference between owning a house and owning a bigger one. Hence, the relative outcome of participating in a lottery can be much higher for a poorer person.

4.3.1 What is the value of a window?
A simple architectural illustration of the effect of marginal utilities relative to benefits received concerns windows. A windowless room can be quite unpleasant, and even a small window makes a considerable difference. A basement room with only a quarter of a square metre of window area is much more pleasant than one with no window; the window can provide a view, ventilation and some natural light. That first bit of window is immensely valuable; indeed, in some office settings

a window is a great status enhancer. However, as the window area becomes progressively larger, the benefit of each extra bit of area declines. When almost all of a wall is window, an additional square metre may not even be perceived by the occupant. The value of the last additional area of window is considerably less than that of the first.

The benefit of increasing the area of glazing for a hypothetical situation is represented graphically in Figure 4.1. The window area is shown on the X (horizontal) axis. The dashed line is the marginal utility offered by each additional bit of window area, and the solid line is the utility offered by the total window area (the numbers are only examples, and are arbitrary). It can be noted that the extra value derived from each additional bit of window area starts high and sinks, approaching zero for very large glazed areas. Hence the total utility curve rises steeply initially and then levels off.

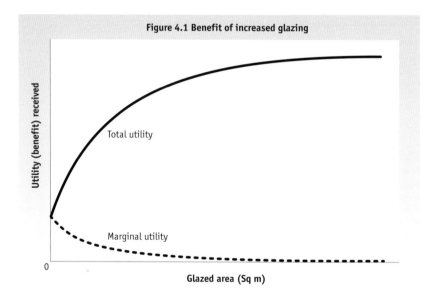

Figure 4.1 Benefit of increased glazing

As with most building matters, on the other side of the consideration is cost (Figure 4.2). Usually the first bit of window is fairly expensive. It has to be on an outside wall and needs a lintel and frame. While after a certain point adding an extra bit of window costs very little (just the extra glass plus a small bit of frame), it still has ongoing costs in terms of heat loss and solar gain, as well as the other life-cycle costs associated with windows, such as cleaning and replacement after the seals fail or someone throws a cricket ball through it. Indeed, after some point, the utility received from an extra bit of area, less cost implications, might become negative, and every additional unit provided actually causes a net loss in value to the project. One utilises resources to create and maintain that extra glass, but does not receive enough benefit in compensation, thereby the window represents an overinvestment and something that may reduce the sustainability of the building.

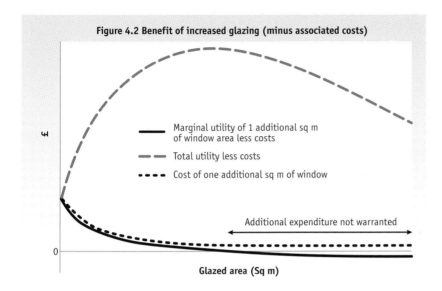

Figure 4.2 Benefit of increased glazing (minus associated costs)

Marginal utility of 1 additional sq m of window area less costs

Total utility less costs

Cost of one additional sq m of window

Additional expenditure not warranted

£

0

Glazed area (Sq m)

This simple example underlines the fact that while some of something might be good, more of it might not be. This example can also be used to illustrate substitution possibilities. Better artificial lighting would be a near-substitute for more glazing area – it provides light, but no view. In an office setting, the morale implications of limiting window area might be compensated for by improved decor or better furniture. As with many sustainability matters, there is often no readily observable market for window area, but the principle still applies. Many benefits are difficult or impossible to quantify, including the enhanced performance that might result from happier users of the space.

Ultimately, some trade-offs have to be made. Perhaps the size of the windows will be related to the costs of better-quality roofing or cladding. At some point the diminishing value at the margin for the larger windows will be perceived as being less than the value that might be realised by expending resources in some other way – but this will vary from person to person and from organisation to organisation.

4.3.2 Adding more features: a calculated example

While the example of the benefits of increased glazing is based primarily on behavioural returns, the concept of marginality can also be important in areas where building outcomes are dominated by physics – such as energy consumption or CO_2 generation.

Consider a hypothetical 5,000 square metre, three-storey office building, located on the edge of a town in northern England. To make the example clearer, we have assumed no mechanical cooling. To enhance energy performance the designer might consider three upgrade possibilities: 1) more insulation, 2) better windows or 3) an improved mechanical system. These might be added to the base design individually, or in any combination.

For illustrative purposes, imagine that the cost of each upgrade possibility was set to give a yield of 7 per cent, based on the energy saved at current energy prices for gas and electricity. That means that if you were to add any one of the alternatives (1, 2 or 3) to the base design, the upgrade would yield 7 per cent – and a decision-maker might therefore be indifferent to the return offered by each alternative.

An energy consumption and cost analysis was undertaken using a readily available building energy planning software tool. Figure 4.3 shows the eight different alternatives (2 × 2 × 2). The results[1] show that although each alternative had the same financial yield when used alone, when used in combination the yield (the return on investment) is lower.

Figure 4.3 Summary of energy costs, savings and yields

Alternatives selected			Energy consumption (£)	Energy savings from base (£)	Capital cost of upgrade(s) selected (£)				Simple yield (%)
Insulation	Windows	Mechanical			Insulation	Windows	Mechanical	Total	
Base	Base	Base	108,345	–	–	–	–	–	–
Upgrade	Base	Base	103,353	4,992	71,314	–	–	71,314	7.0
Base	Upgrade	Base	102,434	5,911	–	84,443	–	84,443	7.0
Upgrade	Base	Upgrade	98,958	9,387	–	–	134,100	134,100	7.0
Upgrade	Base	Upgrade	94,378	13,967	71,314	–	134,100	205,414	6.8
Base	Upgrade	Upgrade	93,760	14,967	–	84,443	134,100	218,543	6.6
Upgrade	Upgrade	Base	99,339	9,006	71,314	84,443		155,757	5.8
Upgrade	Upgrade	Upgrade	91,955	16,390	71,314	84,443	134,100	289,857	5.6

The benefit of adding any specific energy-saving alternative is not a constant, but depends upon the mix of other upgrades also selected. Figure 4.4 shows the impact of two alternative sequences of 'addition', giving the marginal return on investment for additional features.

Figure 4.4 Marginal benefit of adding energy-saving upgrades

Additional sequence: I, W, M			Energy costs	Savings from previous state	Cost of upgrade(s)	additional cost of upgrade(s)	Marginal yields from upgrade(s) (%)
Insulation	Windows	Mechanical					
Base	Base	Base	108,345	–	–	–	–
Upgrade	Base	Base	103,353	4,992	71,314	71,314	7.0
Upgrade	Upgrade	Base	99,339	4,014	155,757	84,443	4.7
Upgrade	Upgrade	Upgrade	91,955	7,384	289,857	134,100	5.5

Additional sequence: W, M, I			Energy costs	Savings from previous state	Cost of upgrade(s)	additional cost of upgrade(s)	Marginal yields from upgrade(s) (%)
Insulation	Windows	Mechanical					
Base	Base	Base	108,345	–	–	–	–
Base	Upgrade	Base	103,353	5,911	84,443	84,443	7.0
Base	Upgrade	Upgrade	99,339	8,674	218,543	134,100	6.5
Upgrade	Upgrade	Upgrade	91,955	1,805	289,857	71,314	2.5

As a building becomes more energy efficient, it becomes increasingly difficult to squeeze out additional energy savings (Figure 4.5). In the example, after energy consumption has been reduced by the first upgrades, there is simply less energy to save, so the returns from additional features are lower.

Figure 4.5 Marginal benefits: energy savings						
Insulation	Windows	Mechanical	Energy used (GJ)	Energy savings from base	Capital cost of upgrade	Return: GJ saved per £1,000
Base	Base	Base	3,385	0	0	–
Upgrade	Base	Base	3,066	319	71,314	43.0
Base	Upgrade	Base	2,924	461	84,443	34.6
Base	Base	Upgrade	3,087	298	134,100	23.0
Upgrade	Base	Upgrade	2,799	586	205,414	13.6
Base	Upgrade	Upgrade	2,676	709	218,543	12.2
Upgrade	Upgrade	Base	2,738	647	155,757	17.6
Upgrade	Upgrade	Upgrade	2,506	879	289,857	8.6

Inevitably, there are additional complications, including the possibility of interactions between the various upgrades. The life cycles of the components vary too. The insulation will effectively last as long as the building and is difficult to install later, especially if it is located under a floor slab; the windows might be subject to future replacement but should last quite a number of decades; substantial elements of the mechanical system will probably require replacement in twenty years or so. For the purposes of this example, component life and the ability to upgrade in future have been ignored (the calculations are simple yields – return divided by cost), but these issues will be considered in subsequent chapters.

4.4 Indifference: Making Trade-offs Between Design Alternatives

Marginal utilities can be understood visually through the structure of indifference curves – how individuals and societies make trade-offs between alternative goods or services. Given limited resources, any individual, organisation or society that is acting rationally (not always a valid assumption) should strive towards a maximisation of total utility (the meeting of needs). This can be achieved by choosing an appropriate mix of goods and services, within the constraints of available resources. This process can be represented graphically by comparing two building attributes. Considering Vitruvius, Alberti and Wotton, these might be delight (some sort of esteem for the visual design of the building) and commodity (its usefulness). Alternatively, the two axes might represent two (of the three) measures of sustainability given on page 7 – perhaps the social and environmental criteria.

Curves can be drawn so that the same total amount of utility is represented at any point on the curve – any mix of social and environmental qualities along the curve would be equally acceptable.

Accordingly, an individual or organisation does not care what point on the curve is ultimately selected or comes to pass – there is an indifference towards the alternative 'bundles' of goods. Figure 4.6 shows three indifference curves. An individual would rather be on a higher curve 'C', as the total level of social and environmental satisfaction is greater. As a solution moves towards the upper right-hand corner of the chart, the level of total satisfaction increases.

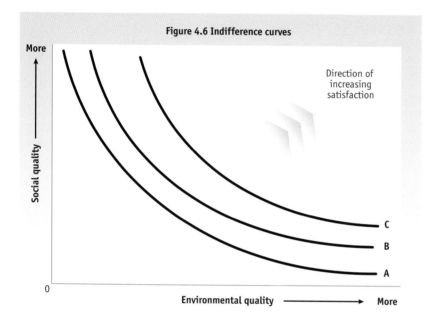

Figure 4.6 Indifference curves

Most indifference curves are (not surprisingly) curved, due to the effects of the changing marginal benefit received from the two goods or services as they change in quantity. Accordingly, the slope varies along the indifference curve; as one possesses more of something, the less gain is felt by having still more of it. For this reason people and organisations are usually willing to trade off more of something they have in abundance in exchange for something they have less of. So the slope of the indifference curve indicates, for very small changes, how much of one good could be sacrificed for a gain of the other. In our example, if the building performs well relative to its social benefits, but is environmentally dubious (top left-hand corner), one might be prepared to give up quite a bit of social performance for just a small amount of environmental improvement. As the trade-offs are made, the amount a decision-maker should 'pay' in the exchange should fall. The exact indifference curve reflects individual preferences – and so the shape and slopes vary from person to person, from market group to market group, and from time to time.

Into this two-dimensional space, various goods or designs (packages of attributes) can be plotted (Figure 4.7). Hence we might have Design A, Design B and so forth. These various alternatives are best seen not as points, but as fuzzy areas encompassing a range of outcomes. In advance, and sometimes even in retrospect, it can be difficult to assess the level of different types of satisfaction that might be obtained from any possible choice. Is a solution really going to be more socially beneficial or environmentally good? The size of this area of fuzziness will vary – the more certain the outcome, the smaller area of fuzziness. Hence there is the possibility of overlapping areas – and the choice may well end up being at least partly based on the decision-maker's propensity to take on risk.

In Figure 4.7, four design alternatives are shown. Given that the upper right-hand corner is the favoured area, Design B is dominated by every other design possibility. The curves suggest that Design A should be preferred to Design D, as it is higher in relation to the indifference curves. They are significantly different, with Design D scoring much higher with respect to social benefit, while Design A is presumably more 'environmental'. Of course each encompasses a range of uncertainty; the orientation of the fuzzy areas shows, for example, that the environmental quality of Design A is more certain than the social outcomes. Design C overlaps with Design A, but the fuzzy area is smaller, implying that the outcome is more certain. Depending on the risk profile of the decision-maker, Design C might be the preferred alternative.

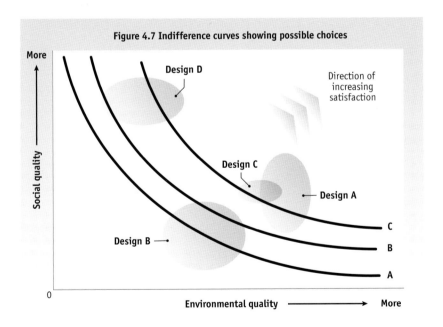

Figure 4.7 Indifference curves showing possible choices

A different person or group of people, perhaps some future generation with other needs and following other design trends, weighing things differently, may come to a different decision (Figure 4.8). In this case, it can be seen that Design D becomes more attractive relative to Design A, not because either design will perform differently, but because of differing individual evaluations. For example, the original curve C has become the dashed curve C' in this alternative evaluation.

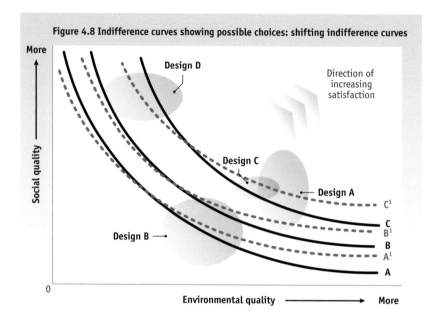

Figure 4.8 Indifference curves showing possible choices: shifting indifference curves

A two-dimensional indifference structure is obviously a simplification of reality – and results from the constraints of two-dimensional paper. A three-axis representation, whereby the different points of indifference would form a curved surface, can also be visualised – perhaps embracing economic, social and environmental performance (or commodity, firmness and delight), but ultimately individuals, companies and societies select from multitudes of different goods and services, each with their own collection of attributes, in their attempt to achieve maximum satisfaction. In reality, our minds make very complex assessments that would require multidimensional representations.

Various observations can be made from this rather simple model of what can be very complex processes, and many can be seen in two-dimensional examples.

Different people and organisations will make different trade-offs – for example, some people would like more space – so to get the house of their dreams they will be more willing than most people to reduce their consumption in other areas. Others may prefer holidays in the sun.

Indifference curves are downward-sloping and curved. As the quantity of one commodity is reduced, the other has to be increased relatively more to compensate. The relative 'value' of the different goods or services is not constant.

Changes in available resources may alter the optimal selection. With most goods and services, as income increases, higher-quality goods and services will be preferred, and fewer inferior goods will be chosen. Increasing the income of an individual will typically move the curve in a more or less parallel fashion to the right. However, increasing affluence can cause the desirability of some goods to drop (for example, wealthy people tend to use little imitation stone cladding).

Some special cases exist, in particular when goods can only be used together, where no additional utility/satisfaction results from having more of one commodity without having more of the other. Buildings in cold climates have little value without fuel. This tends to create 'L'-shaped curves.

People, organisations and societies have resource limitations – something that governments discover from time to time when they exceed them. This limits the move to the upper right-hand corner of the chart. The combination of the straight resource (or budget) limit line and the indifference curves creates a point of tangency – at which, given budget realities, a maximum level of utility will be reached.

4.5 Indifference and Sustainability

Indifference curves may help us understand the trade-offs that have to be made when making decisions in sustainable design. One of the more difficult compromises is between different measures of sustainability, because most decisions encompass economic, environmental and social/cultural concerns. While cost implications, both initial and longer term, can be drawn to a single monetary measure, there are many kinds of environmental measures, some of which are exceedingly difficult to quantify. One might use such graphs to conceptualise different aspects of environmental performance on the different axes – perhaps trading off emissions-free electricity against species protection.

While plotting the different alternatives might be reasonably objective, relating the significance of the different attributes and how to trade them off will always be heavily influenced by the individual or group doing the analysis.

4.6 What Might This Mean to the Designer/Decision-maker?

The concept of marginal utilities can offer interesting insights into how design decisions might be undertaken. It should be recognised that a designer or manager should not just keep adding features in an attempt to be 'green' or 'sustainable'. The entire package is important, not just the constituent elements. Balance is required to avoid either overinvestment or underinvestment. This explains why a building design that may already include a high level of insulation, triple-glazed windows, a natural ventilation system and an air-to-air heat exchanger probably does not warrant a ground-source heat pump system, which could well constitute overinvestment – using resources that could be put to better use in other ways. Such buildings have been built, but the additional expenditure may have been warranted only through the prestige and marketing value resulting from a high-level building certification.

Marginal utilities underline the inherent problem in government policies aimed at creating 'zero carbon' houses, or other such dramatic policies as a way of achieving CO_2 reduction or energy efficiency, other than as demonstration projects. Presumably it would sometimes be better to spread the always-limited resources among the many poorly performing buildings in the existing stock (and other opportunities, such as in transportation), rather than offering substantial incentives or policies to create a few high-performing new buildings.

For the designer or manager seeking to create buildings that contribute to sustainability, the concept of marginality is very important. Decisions, whether made using paper and pencil, a spreadsheet, or through some other mysterious cognitive process, must all respect this fundamental reality. Endlessly adding increasing numbers of features billed as 'green' may be fundamentally unproductive.

4: Notes

1   It might be noted that the energy consumption is a mixture of gas and electricity as allocated by the underlying programme, so the energy cost does not immediately correspond to the absolute energy consumption, nor does the energy consumption directly correspond to the amount of CO_2 generated by this hypothetical building.

Time and Risk

5

5.1 Time Preference: the Long Term and the Short Term

Sustainable design decisions are those we take now that have long-term benefits.
Sometimes the path through time is simple: we plant a tree today and it grows
and (hopefully) gives pleasure for a hundred years or so; or we decide not to
chop down a tree, yielding a similar benefit. Other cases are more complex: rather
than refurbishing a fifty-year-old office block we may decide to knock it down and
build a new replacement, in the belief that over time the new building's improved
thermal performance and higher rental returns will more than compensate for the
extra embodied emissions and money invested in the redevelopment. The decision
about whether to plant or cut down a tree could probably be made with a few
minutes' clear thinking, but the decision about whether to refurbish or redevelop
requires more serious analysis, with numbers attached.

A refurbishment/redevelopment problem can be divided into two parts – data
and time. First, we need data estimates for the construction cost and resource
flows (including embodied energy and emissions) of the refurbishment and
redevelopment, and the flow of annual incomes and resources for the two

alternatives after they are occupied. Then we have to balance the costs and benefits that arise at different times to establish whether in the long run it is more sustainable to refurbish the old office or build a new one.

How does one decide between alternatives that (may) offer benefits and incur costs at different times, and with different degrees of certainty?

One fundamental issue is time preference. We often have strong preferences about whether an event happens now or in the future (if only things could happen differently in the past!). In some situations we are unwilling to wait. For example, if we're really hungry, baked beans on toast now might be preferable to a three-course meal in four hours' time. In other situations we're delighted to put things off, perhaps deferring a client meeting from today until tomorrow so that the necessary drawings can be completed first.

In general we prefer good things to happen now or soon, and prefer bad things to be delayed as long as possible. Suppose that a good event, say a £1 million one-off windfall or a saving of 100 tonnes of CO_2 emissions, could take place at one of four alternative points in time: 1) this year, 2) next year, 3) in five years' time or 4) in twenty years' time. How would you rank the alternatives? Most people would rank them 1–2–3–4 in order of preference. And what about a bad event, say a £1 million one-off bill or 100-tonne surge of CO_2 emissions, at the same four alternative points in time? Most people would rank them 4–3–2–1 in order of preference. We need to turn this observation into a tool for evaluating alternatives.

5.2 The Basics of Discounting

One of the most commonly used ways of analysing or simply conceptualising the time dimension of alternatives with costs or returns at different points in time is discounting. Discounting is the engine of many decision-making tools, regardless of what they are called or what their acronyms spell out. Net Present Value (NPV), Discounted Cash Flow (DCF), Whole-Life Costing (WLC), Life-Cycle Costing (LCC) and Cost–Benefit Analysis (CBA) are all discounting-based tools. Although discounting is widely used, it is often poorly understood and regarded as a black box that can't be opened. This is wrong: don't be put off by the attendant acronyms and any apparent complexity – it is simpler than you think. We're going to open the box and show that its contents are basically a matter of common sense.

The power of discounting derives from one crucial simplifying assumption, which says that the impact of deferring an event by one year can be captured as a percentage value, called the discount rate. For example, if the discount rate is, say, 5 per cent, then today's evaluation of the consequence of an event taking place in one year's time is correspondingly less than the value of the same event if it took place today. The same discount rate applies for the following years as well, so that today's value of the payment/charge taking place in two years' time would be less. The discounting process is repeated in exactly the same way for every year into the future.

It is important to note that discounting doesn't actually make the event smaller when it eventually happens. It simply means that from today's perspective, future events count for less than current events. There are a number of reasons why this is so, and it is this collection of elements that make up discount rates.

Try to imagine trading between current events and future events. How much would you pay for £1 million paid in five years' time if your discount rate was 5 per cent? The discounting formula tells us that the present value equivalent of £1 million in five years is £783,526, so if the simplifying assumption of discounting were correct (and the right discount rate is being used) you would willingly exchange £1 million after five years for £783,526 now. Discounting applies to both good events and bad events – the pleasure value of good events is reduced if they are delayed, and the pain value of undesirable events is also reduced. It is very much like borrowing from or lending to a bank with interest.

To deal with the time dimension in the refurbishment/redevelopment problem described in Section 5.1, all the data estimates for the construction cost and embodied emissions, and the flow of annual incomes and emissions and social benefits for the two alternatives, can be discounted from the year in which they occur back to the present, giving a stack of present value equivalents. They are all added up for the refurbishment and redevelopment alternatives, and the alternative with a greater total of present values is preferred. This is the NPV of the alternatives.

While people tend to see discounting as being about money, in reality monetary measures act only as proxies for the real goods, services and other resources that we desire to support our well-being, now and in the future. Money is just a convenient common measure.

The mathematics of basic discounting are quite simple. You just apply the same discount rate over and over again off into the future. It can be done easily on a spreadsheet. The problem isn't the mathematics but the choice of the discount rate, obtaining meaningful data and interpreting the output. Outcomes are highly sensitive to the discount rate, especially as you move far into the future when the discount rate is applied many times. If you use inappropriate discount rates, any decisions that you base on them may be useless.

5.3 The Logic of Discounting

Discounting alone is not the ultimate tool to deal with events that occur at different times, and needs to be used with caution. As we will see, in many cases add-on techniques are required, and the results should always be subjected to scrutiny. Environmental specialists are often very unwilling to apply discounting to environmental impacts. However, it is necessary to reflect upon at least some of the aspects of time and uncertainty when assessing project alternatives, otherwise the results can be illogical. Discounting is the usual method when dealing with time and risk, although we will consider other approaches later.

It is worth noting that a person or organisation with a very pronounced time preference, strongly favouring present over future pleasure and strongly preferring to defer the incidence of pain, will have a high discount rate. Someone else, less impatient for pleasure and less desperate to delay pain, would have a lower discount rate. Put another way, a high discount rate equates to a short time horizon (or short-sighted decision-making); a low discount rate equates to a long-term view.

A zero discount rate is mathematically possible, and is sometimes advocated for environmental analysis – often on moral grounds. This means that whatever happens in the future is given exactly the same weight as if it were to happen today. The economist Paul Samuelson (1915–2009), a proponent of discounting, pointed out that this would lead to irrational decisions, using the example of railway lines through hilly terrain.[1] As the track becomes flatter, trains encounter fewer energy-sapping gradients and operate more economically, but flatter lines are more expensive to build. Engineers compromise between the initial cost of construction and the operating costs of fuel and engine/brake wear and tear that accrue in each year of operation. The trade-off depends on the discount rate. With a high discount rate the engineers will design a cheap line with steep gradients, and although there will be high running costs in every year of operation, they will be discounted at a high rate. If the discount rate is low they may decide to build a more expensive line with flatter gradients, with the extra construction cost being offset by many years of running cost savings. But with a zero discount rate, the engineers would spend unlimited amounts of money to build a perfectly flat line, because in the long run, perhaps over hundreds (or thousands) of years, the accumulated undiscounted operating cost savings, no matter how small, would eventually offset the construction cost. The logical result would be that the generation building the railway line would invest massive resources in exchange for benefits that might accrue hundreds of years in the future – a classic case of irrational overinvestment, if for no other reason than transport needs change, and rail lines are sometimes abandoned.

One point about the choice of discount rate can be made at once: there is no 'standard' rate. A widely used discount rate in the UK is 3.5 per cent, because the Treasury has calculated that this is appropriate for public sector projects funded by the government. The government is a big investor so it has specific characteristics that do not apply to other decision-makers (not least its size), and others should not blindly use the government's discount rate.

5.4 The Components of the Discount Rate

Discount rates are composed of a collection of elements that are added together to create an overall rate. In the management world discount rates are usually set based on observation of market behaviour. A company can create a Weighted Average Cost of Capital (WACC) based on how the market values its stock and debt instruments. In the property world, especially when environmental and social/cultural matters arise, more insight is required.

5.4.1 Pure time preference

Pure time preference is a matter of personal or organisational psychology. It varies from individual to individual and from organisation to organisation. Typical figures for groups of people have been established in surveys. In the UK the average pure time preference for the population as a whole is believed to be about 1.5 per cent per year.[2]

5.4.2 Expectation of increasing well-being

The discount rate establishes today's equivalent of a cost or benefit arising in the future: if we're expecting to be better off in the future this must be taken into account in establishing the equivalence. Think about when you were a starving student – things that were luxuries then (perhaps a restaurant meal or a used car) would be quite unexceptional for most middle-aged architects. Perhaps the ratio of incomes might be 10:1; then £100 to the student would be equivalent to £1,000 to the prosperous architect; and from the student's perspective, today's equivalent of a cost or benefit of £1,000 arising in the future would be discounted to £100 today due to the expectation of rising wealth. A student would sacrifice relatively little for a meal to be received five years in the future, so the value of that future meal must be discounted substantially. This is the future wealth component of the discount rate.

Just as with pure time preference, the future wealth component of the discount rate varies from individual to individual and from organisation to organisation, depending on expectations. For a student looking forward to a successful professional career, the future wealth component would be high. Few young people save for the future because they assume that they will become wealthier: a severe sacrifice of present consumption would generate relatively pitiful future benefits. But for someone approaching retirement with a modest pension there might be an expectation of falling wealth, and a small or negative future wealth component of the discount rate. Thus housing developers have commented that buyers aged over seventy will happily pay for energy-saving features, but they're often of little interest to first-time buyers.

For the UK as a whole the government expects a long-run rate of increase in national wealth of 2 per cent, and this 2 per cent figure is used as the future wealth component of the government's discount rate. But a different future wealth component, higher or lower, would be more appropriate for many non-government investors, as well as other societies in which growth might be higher (developing economies) or lower (stumbling economies with economic difficulties).

5.4.3 Uncertainty

When dealing with future events, some uncertainty is always present. When establishing today's equivalent of costs or benefits arising in the future, we must take account of the uncertainty surrounding these costs or benefits. Uncertainty (or risk) is often the most important part of a discount rate, because it can differ substantially from situation to situation and from time to time.

People prefer greater certainty, and attach decreasing value to events as their uncertainty rises. Therefore most discount rates have a component reflecting uncertainty, called the risk premium, which rises with the amount of uncertainty. A risk-free investment has a zero risk premium. The traditional example cited as a risk-free investment was a US government bond, but even here a risk premium can creep in.

In building projects, some costs might be known quite well – perhaps the cost of land and the various charges levied by government. Others are less certain, such as the construction cost, and still others become increasingly uncertain as a function of time, such as the level of rent that might be received in ten years' time. Most of the numbers used in a project analysis are forecasts of expected future events, and generally there is more uncertainty associated with events further away in time. This is neatly accommodated by the risk premium, which has the effect of decreasing the relative importance of distant events in an analysis.

There are many sources of uncertainty. One is technological change. New components and systems are always being developed, and while there is no assurance that there will be new products which give an increase in performance, this is very often the case. Although innovation cannot be predicted, its potential impact should be recognised, particularly when dealing with emerging technologies for energy-saving or CO_2 reductions.

The UK government includes a small risk premium in the discount rate it specifies for public sector investment. This is because the government has such a wide range of investment projects that the risks in individual projects balance each other out. The small risk premium is to cover the possibility that the British population may not be around to enjoy the benefits of investments – having been wiped out or significantly disrupted by some catastrophic event such as an epidemic, meteor strike or thermonuclear war. Individuals and corporations will add substantial amounts to this basic risk premium to reflect private risks – a shopping centre developer would add a considerable amount to allow for the fact that the market for his shopping centre might not develop as expected, for example.

5.4.4 Inflation

Inflation only applies to discounting when costs and benefits are expressed in monetary terms. If they are expressed in terms of CO_2 emissions or other physical units you do not have to worry about inflation.

Inflation is the process whereby prices for exactly the same goods and services increase over time. With deflation, which is much rarer and even more unwelcome, prices for exactly the same goods and services decrease over time. It might be a surprise to realise that inflation has not always existed. There was very little inflation in the Middle Ages because of the constrained supply of silver and gold coin, but it appeared along with the flow of gold and silver from Latin America.

A brief consideration of the ever-changing rates of inflation clearly indicates that any long-term inflation projection (following the kind of time frame in which buildings exist) will inevitably be wrong. While inflation rates during the first decade of the twenty-first century have ranged in the 1 to 2 per cent range, throughout the late 1970s and early 1980s rates ranged up to over 20 per cent per year.

Inflation can be troublesome for money-based discounting, but the important thing is to be consistent: either include inflation in both the cost estimates AND the discount rate, or eliminate inflation from both the cost estimates AND the discount rate – never mix these approaches. Because inflation cannot be predicted, it is usually best to eliminate it. Then future estimates are made in what are called 'real prices'; that is, the prices that would apply in the future if inflation stopped dead today, as if goods and services would have exactly the same cost in the future as they have today. Since the purpose of discounting is to transform future costs and benefits into today's terms, you will always end up with a present value expressed in today's value of money, so there's no problem in using today's value of money throughout the analysis.

This is one of the differences between discounting and bank interest rates, which always include some allowance for inflation.

5.5 What Do I Do Next?

Architects are constantly dealing with their clients' time preferences, but these are rarely expressed as a single-number discount rate. If you ask a client, even in an organisation with a substantial building programme, what discount rate they would like to use for a project you're very unlikely to get a worthwhile answer, or indeed any answer at all. But you can still probe the issue of time preference with respect to the four underlying factors: the client's pure time preference, risk attitudes, expectation of future wealth and the uncertainty surrounding the project. This should help determine whether for a given client and project a long time horizon is appropriate, with high initial investment justified by savings to be made over a long period (this would be a low discount rate situation) or whether a shorter time horizon is more appropriate, with a lower initial investment and higher running costs (a high discount rate situation).

A sustainability-focused architect would probably prefer to work on projects with a long time horizon, i.e. a low discount rate, but it is inappropriate to impose a low discount rate on a client and project when a high discount rate would be more suitable. This is likely to lead to conflicts in meetings, and poor and unsustainable decisions.

Most architects avoid undertaking a discounting-based analysis, even though it is technically quite easy, often leaving it to a quantity surveyor or cost or investment consultant. The unfortunate result is that such an analysis tends not to be used as a decision-making tool. Perhaps this is because there are assumptions that discounting analyses are to be applied to fully designed projects, and that there are vast amounts of data required, together with specialist expertise. Yet, as a decision-making tool, even back-of-the-envelope calculations to explore design decisions as they are made, using whatever data is available, can help in making more sustainable decisions.

5: Notes

1   Paul Samuelson (1967), *Economics* (7th edition), New York: McGraw-Hill, p. 577

2   HM Treasury (2003), *The Green Book: Appraisal and Evaluation in Central Government*, London: HM Treasury

Quantification: Undertaking an Analysis

6

In the quest to become an expert decision-maker, an understanding of some basic quantitative decision-making tools is necessary. Although many decision-makers will never assemble the numbers or build a spreadsheet to make the calculations, just knowing about the logic and methods, the contexts in which they function, and their limitations, will assist in assessing alternatives and making real-world decisions of all kinds.

The mathematical techniques of whole-life costing analysis are dealt with at great length in some texts, but the reality is that when considering any form of decision-making, sustainable or otherwise, specific techniques can be the easiest part of the entire exercise. Framing the project, developing alternatives, collecting data, understanding what the data means and interpreting output and how to respond to it are much more difficult. Decision-making is about much more than putting numbers in a black box, turning the handle, and having the answer appear.

One essential objective of decision-making is simple. When individuals, organisations and societies make decisions, they inescapably ask: 'By how much should we defer consumption in the present to receive an increased ability to

consume in the future, given all the characteristics of such a trade-off?' This
not only includes matters of business, but also encompasses the fundamental
definitions of sustainability – we do certain things in the short term in anticipation
that we, individually or collectively, will be somehow better off in the future. From
an economic perspective, this means not buying something today – perhaps a more
expensive car – and putting some money away, perhaps to be used in retirement.
It can also mean turning the temperature down in one's house to save some fuel
that might be used a century or more in the future. And of course, we make many
trade-offs between the economic, environmental and social/cultural spheres. For
instance, we collectively, through government, fund sports facilities, concert halls
and art galleries that we anticipate will improve society as a whole in the longer
term. As individuals we will forego consumption now and pay for our childrens'
education, something that may, or may not, have some benefit in the future – and
probably not directly to us.

Sustainability questions can make numerical analysis very difficult – but not
impossible. In reviewing the basic techniques, though, it is best to start by simply
using money as a measure and then adding on the additional layers.

6.1 Ms Napier Looks at Flooring

Catharine Napier is a staff architect for a major department store chain. She is
working on a new outlet, and is considering different flooring materials for the
sales areas. The problem is not a new one. Indeed, the company has never quite
determined what the best alternative is. They tend to use vinyl tile, because it is
reasonably cheap and durable, but probably mostly because it has always been done
that way. Different people in the organisation have their various opinions – the
department managers like certain finishes because they think they make the space
more appealing and so lead to greater sales. The corporate finance people always
worry about cost effectiveness, and the facility's operating staff like low-maintenance
materials. Catharine Napier's own department does not want to disrupt operations
too often, so likes long-life materials – but not too long, because fashions do change,
and sometimes materials are replaced when a new 'look' is required rather than
because they are worn out. She really would like to solve the problem – not just
for this project, but perhaps hoping to give rise to a wider policy on flooring for
future projects. There are quite a few issues and opinions, and even the costs of
maintenance are not entirely clear, but the department head thinks it might be
worthwhile for Ms Napier to work on the problem and see what might unfold.

The first thing to do is to collect some data. As is typical in the building world,
things are not straightforward. The cost of maintenance seems to vary among the
company's various locations; however, Ms Napier can develop some sense of what
maintenance costs the various alternatives will incur. The store managers have also
given her a range of opinions about how their sales might change depending upon
the flooring chosen. They all suspect that vinyl tile is probably not the best thing,

so have been happy to give her whatever information they have. She also used the profit margin of the company to convert the sales per square metre into some notion of additional profit that might result.

Even given the mixed quality of the data, Ms Napier manages to assemble a table of likely alternatives to use as the basis for her analysis (Figure 6.1).

Figure 6.1 Flooring materials, costs, life expectancies and implications

Flooring materials	Vinyl tile	Quality carpet	Cheap carpet	Rubber tile	Terrazzo
Initial cost / (£/Sq m)	54.37	63.25	40.25	79.89	171.35
Expected physical life (years)	12	8	5	20	100
Annual maintenance costs (£/Sq m)	6.03	5.29	5.88	5.13	5.13
Appearance (1–10: 10 is best)	5	10	7	7	7
Possible additional sales-based profits (£/Sq m)	0.00	3.50	3.00	2.50	2.50

She might organise the data to enable a comparison of the costs and benefits offered by the alternatives relative to the vinyl tile currently being used (Figure 6.2).

Figure 6.2 Reorganised data

	Vinyl tile	Quality carpet	Cheap carpet	Rubber tile	Terrazzo
Extra initial cost (£)	0.00	8.88	-14.12	25.52	116.98
Maintenance savings (£)	0.00	0.74	0.15	0.90	0.90
Additional profit (£)	0.00	3.50	3.00	2.50	2.50
Total Benefit relative to vinyl tile (£)	0.00	4.24	3.15	3.40	3.40

Simply inspecting the results reveals something that decades of managers may have missed or ignored: low-priced carpeting is not only cheaper to install than the vinyl tile, but has lower maintenance costs and might be expected to increase sales and overall profit. However, Ms Napier notes the short life, and suspects that is one reason why carpeting has been avoided. Recarpeting involves considerable disruption, and presumably lost sales. Moreover she regards the increase in sales from cheap carpeting as suspect, as towards the end of its life the carpet will tend to become less attractive.

6.2 Simple Payback and Simple Yields

Payback methods, described in various ways (cap rate, year's purchase …), are very simple, basic ways of comparing alternatives, and remain popular in the property industry because property investments often involve a one-time expenditure of resources (investment), and then a long-term, more or less constant return of benefits (savings or returns). They also offer a very simple way to start an analysis – just don't take them too far. A typical decision asks whether the benefits justify some extra initial expenditure, perhaps a higher-specification building component, and if there are several alternatives, which of them might be the best. This is the way many decisions present themselves to architects.

A simple but rudimentary method is to divide the extra costs by the extra annual benefits. This gives a 'payback period' in years; short payback periods are preferred. In these examples, inflows and outflows of resources have been used. Formulae such as these usually use money – inflows of money (benefits) and outflows (costs) – however, money only acts as a proxy for other things that might be of use.

The formula is:

$$\text{Payback period} = \frac{\text{Initial outflow of resources (costs)}}{\text{Annual inflow of resources (benefits)}}$$

or

$$f = \frac{C}{b}$$

where f is the payback period, C is the initial outflow of resources (capital cost) and b is the annual inflow.

Using this formula, Ms Napier can calculate paybacks for the various alternatives. For example, for the rubber tile, if there is an annual benefit of £3.40 per square metre, and an initial additional cost of £25.52 per square metre:

$$f = \frac{25.52}{3.40}$$

$$f = 7.5 \text{ years}$$

This payback might be compared with the paybacks of the other alternatives – the lower the number, the better.

This can be reversed to express a simple return on investment or yield (y) for the additional expenditure:

Simple yield $\quad = \quad \dfrac{\text{Annual inflow of resources (benefits)}}{\text{Initial outflow of resources (costs)}}$

or $\qquad\qquad y \quad = \quad \dfrac{b}{C}$

For example, for the rubber tile:

$$y \quad = \quad \dfrac{3.40}{25.52}$$

$$y \quad = \quad 13.3 \text{ per cent}$$

As these approaches are effectively identical, they will always identify the same preferred projects – expressed as the lowest payback period (in years) or the highest yield.

Ms Napier immediately sees that these are very limited tools, because they do not take into account how long each alternative will last. The carpet alternatives have quite limited life expectancies, especially when considered against terrazzo, which might be expected to last as long as the building. Payback and simple yields ignore longer-term events and, while they can be useful, caution must be exercised – they are incomplete decision-making tools.

6.3 The Basics of Whole-life Costing: Dealing with Time and Uncertainty

A better way is to look at the entire stream of benefits expected from possible alternatives over the project or component life, or over a specific study period. One way might be to simply add them up and compare them with the original costs. The reason this makes limited sense is because of time preference and risk (think of Samuelson's railway – see Section 5.3). The problem is how to adjust for the effects of time and uncertainty. The commonly accepted and almost inevitable approach when comparing future resource flows to present resource flows is to adjust them by deducting a given amount. While other strategies are possible, the level of complexity can increase dramatically.

Uncertainty or Risk?

People frequently use the terms 'uncertainty' and 'risk' interchangeably, often differentiating them on the basis that risk carries more negative connotations, but economists tend to follow the definition given by Frank Knight in his book *Risk, Uncertainty, and Profit* (1921).

Risk occurs when future events are associated with a measurable probability. A simple example is a roulette wheel. While the result of the next spin is unknown, the probability of it being a '12' or 'red' *is* known.

Uncertainty occurs when the likelihood of future events is indefinite or cannot be calculated. While we sometimes make predictions about likelihood, they can seldom be quantified, especially when an event may only occur once, such as the need to expand a building or the possibility that life on Earth will be destroyed within the next century.

In that uncertainty does not usually imply a judgement about whether the result will be fortunate or unfortunate, it will generally be used herein. Risk, when we use it in this book, will tend to imply a situation that apparently has greater negative possibilities.

… the truth is, there are things we know, and we know we know them – the known knowns. There are things we know that we don't know – the known unknowns. And there are unknown unknowns: the things we do not yet know that we do not know.
Donald Rumsfeld, US Secretary of Defense, 1975–7, 2001–6

We saw in Chapter 5 how time preference, expectations of increasing wealth and uncertainty lead to the convention of discounting, whereby future estimated expected costs or benefits are scaled down. A high discount rate greatly reduces the weight attached to what happens in the future; a low discount rate gives more weight to future resource flows. However, even using low discount rates events in the distant future tend to disappear.

Discounting methods are a valuable tool in the building world to analyse alternatives that have different initial costs, different life expectancies and different ongoing costs or benefits. As in the flooring example given (see Section 6.1), often an alternative with a higher initial cost will yield greater benefits in the future – which in monetary terms can be recognised as 'expected cash flows'. Expected cash flows – in this case all costs – for each alternative, perhaps for cheap and quality carpet, can be visualised as a graph of cumulative costs (Figure 6.3). There is an initial expenditure, and then ongoing maintenance and replacement costs over time. At some point the lines will cross if the alternatives last long enough.

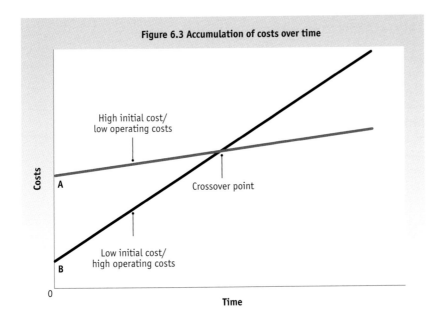

Figure 6.3 Accumulation of costs over time

The recognition of time and uncertainty means that the returns received in the future are of lesser value in the present. Different graphs appear, whereby the time taken for the lines to cross is longer, and in a discounting context, the higher the discount rate, the longer it takes. If the lines do not cross within the time period of the study, or the life of the alternatives, the low-cost alternative is preferred (Figure 6.4).

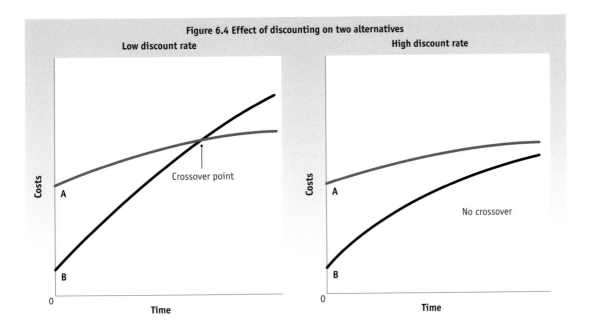

Figure 6.4 Effect of discounting on two alternatives

The discount rate used can range widely. While the UK Treasury uses a discount rate of 3.5 per cent per year for money-based appraisal of public sector investment, private sector entrepreneurs sometimes use rates of 30 per cent or more. The main difference is in the 'risk premium'. While risk is a major factor for individual investors, governments can ignore much risk for two reasons. First, they hold a large portfolio of projects so tend to be less sensitive to the risk associated with any one of them, and second, the risk is spread among millions of taxpayers, each of whom is relatively insensitive to losing a pound or two.

In discounting, one might either numerically or conceptually accept a resource flow in the present at its face value. For events occurring in the future the usual formula is:

$$V_0 = \frac{V_i}{(1+r)^t}$$

Where v_i is the amount of the future flow of resources, r is the discount rate and t is the number of periods (in the case of the built environment typically expressed in terms of years), and v_0, the 'present value', is the equivalent value of that cash flow at time t = 0 (the start of the project).

Hence, if there is a one-time flow of resources ten years in the future valued at £1,000, and it is discounted by the Treasury *Green Book* social time preference rate of 3.5 per cent, it equates to an amount received today of:

$$V_0 = \frac{1,000}{(1+.035)^{10}}$$

$$V_0 = \frac{1,000}{(1.035)^{10}}$$

$$V_0 = \frac{1,000}{1.4106}$$

$$V_0 = £708.91$$

This is done for the resource flows in each period. A number of shortcuts exist, in particular the return for perpetuities – situations when the benefit continues effectively forever – essentially the simple yield calculation seen above.

$$V_0 = \frac{v}{r}$$

So, if the £1,000 continues annually forever:

$$V_0 = \frac{1{,}000}{.035}$$

$$V_0 = \frac{1{,}000}{.035}$$

$$V_0 = £\,28{,}571.43$$

Hence, it would be worth spending the £28,600 to get that long-term flow of benefits. If a higher discount rate were chosen, you would pay less.

The overall result, the NPV (Net Present Value), is the sum of the individual discounted cash flows, including the original expenditure (a negative amount), and may be either positive or negative. A positive NPV suggests that an alternative has some merit, while a negative number suggests that proceeding with it probably represents overinvestment. A decision-maker should select the alternative, or set of alternatives, that offers the highest NPV.

Dealing with Inflation

In this case, inflation has been ignored, and real prices have been used as if there was no inflation (instead of nominal numbers including some inflation assumption).

Of course, if there is a significant expenditure point five years in the future, prudence would suggest offering some sort of increase so blame for underestimation does not fall too heavily on the analyst.

Any discount rate used with real numbers needs to be real as well – again without inflation. Market interest rates and companies' 'weighted average cost of capital' do include a factor which is the market's belief about how future inflation might erode the value of money.

Catharine Napier might lay out the costs and benefits of the flooring alternatives on time lines. In this case, she is attempting to minimise the costs of the flooring solution. She will still prefer the alternative with the highest number, even though the numbers (mostly) represent costs, so the answer will be a negative number.

If she plots the costs of the various alternatives she will be able to see which might be preferred. Figure 6.5 compares vinyl tile and quality carpeting, when an 8 per cent discount rate is used.

Figure 6.5 Comparing two alternatives (first 12 years shown of a 35-year analysis)

Discount rate: 8%   Figures in £   Outflows shown in brackets

Years	0	1	2	3	4	5	6	7	8	9	10	11	12
Vinyl tile													
Installation	(54.37)										Tile replacement:		(54.37)
Maintenance	(6.03)	(6.03)	(6.03)	(6.03)	(6.03)	(6.03)	(6.03)	(6.03)	(6.03)	(6.03)	(6.03)	(6.03)	(6.03)
Additional profit	0.00	0.00	0.00	0.00	0.00	0.00	0.00	0.00	0.00	0.00	0.00	0.00	0.00
Total annual flow	(60.40)	(6.03)	(6.03)	(6.03)	(6.03)	(6.03)	(6.03)	(6.03)	(6.03)	(6.03)	(6.03)	(6.03)	(60.40)
Present value of annual flow	(60.40)	(5.59)	(5.17)	(4.79)	(4.43)	(4.11)	(3.80)	(3.52)	(3.26)	(3.02)	(2.79)	(2.59)	(23.99)
Net present value	(160.87)												
Quality carpet													
Installation	(63.25)					Carpet replacement:			(63.25)				
Maintenance	(5.29)	(5.29)	(5.29)	(5.29)	(5.29)	(5.29)	(5.29)	(5.29)	(5.29)	(5.29)	(5.29)	(5.29)	(5.29)
Additional profit	3.50	3.50	3.50	3.50	3.50	3.50	3.50	3.50	3.50	3.50	3.50	3.50	3.50
Total annual flow	(65.04)	(1.79)	(1.79)	(1.79)	(1.79)	(1.79)	(1.79)	(1.79)	(65.04)	(1.79)	(1.79)	(1.79)	(1.79)
Present value of annual flow	(65.04)	(1.66)	(1.53)	(1.42)	(1.32)	(1.22)	(1.13)	(1.04)	(35.14)	(0.90)	(0.83)	(0.77)	(0.71)
Net present value	(148.51)												

Doing this analysis for all of the alternatives shows that, from a strictly monetary perspective, the rubber tile is preferred, having the highest net present value (lowest cost in this situation) over the thirty-five-year analysis period. It is followed by the two forms of carpeting, which have very similar net present values.

Of course the answer depends upon the discount rate used. A higher discount rate would bias the analysis in favour of solutions with a lower first cost (Figure 6.6). In this case, carpets become more attractive. A high discount rate tends to focus on short-term matters – in particular a low initial cost. A low discount rate would assign relatively more value to future events, so the long-life terrazzo would appear relatively better. This shows why the world is so often viewed from a discounted perspective – otherwise, as the discount rate approaches zero, it becomes appropriate to pay infinite amounts for value to be received far in the distant future. Terrazzo might be expected to still be performing in a century or so, but the building may no longer be standing, or may have a different use, or terrazzo may have become unfashionable. In those cases, the far-off benefits may not be received.

Figure 6.6 Net present value (NPV) comparison – different discount rates					
	Vinyl tile	Quality carpet	Cheap carpet	Rubber tile	Terrazzo
NPV at 3.5%	-240.84	-213.05	-213.89	-175.22	-226.53
NPV at 5%	-206.31	-185.75	-184.22	-155.65	-217.00
NPV at 8%	-160.87	-148.51	-143.58	-130.28	-204.60
NPV at 12%	-127.25	-119.70	-111.76	-112.28	-195.46
NPV at 15%	-84.08	-80.85	-66.67	-82.52	-183.10

Note that numbers are costs (negatives), and the objective is to minimise cost.

Some texts spend considerable time on the intricacies of the calculations, but for the building practitioner it is more important to simply grasp the concepts, to be able to mathematically undertake simple discounting calculations, to be able to understand and utilise the results and, inevitably, to be able to criticise an analysis done by someone else.

Whole-life Costing Formulae

While this field can become quite complex, there are really only two formulae to remember when undertaking conventional whole-life costing.

The value of a cash flow in perpetuity:

$$V_0 = \frac{V}{r}$$

and the value of a one-time cash flow at some point in the future:

$$V_0 = \frac{V_i}{(1 + r)^t}$$

where:
- V_i is the amount of the future flow of resources;
- r is the discount rate and t is the number of periods (in the case of the built environment typically expressed in terms of years);
- V_0, the 'present value', is the equivalent value of that cash flow at time $t = 0$ (the start of the project).

6.4 Internal Rate of Return

Practitioners may encounter the Internal Rate of Return (IRR). This is nothing more than a reversal of the discounting process, and answers the question 'given the positive and negative resource flows (costs and benefits incurred and received over time), what is the discount rate (rate of return) at which the net present value is zero?' The logic is simple – when the net present value is zero the time-adjusted amounts of costs and returns balance. Each project alternative will offer a rate of return (which is an improvement over the yields expressed by simple return calculations) and the resulting percentages can be compared.

The internal rate of return can be a convenient measure – as long as it is used with caution, as it has a number of limitations, including the following two most important ones.

- **Comparison of alternatives may be difficult.** The process of undertaking a whole-life costing analysis requires the setting of discount rates and some understanding of associated risks. Different alternatives may have differing likelihoods of achieving favourable outcomes. With an IRR measure, the tacit assumption is that the decision-maker weighs the risks of the various alternatives some other way. Therefore, it is all too easy to be attracted to the highest rate of return, which might be a riskier project, and the extra return may not be enough to compensate for the extra risk.

- **The type of analysis can produce multiple solutions.** This will not occur for most property-based decisions because of their simplicity. Typically one makes an initial expenditure and receives the benefits over time. However, if there are major resource expenditures in the future, perhaps for refurbishment or disposal, or the dismantling of a nuclear power station or site remediation, this can be an issue.

6.5 What Do the Results Mean?

Catharine Napier, as a prudent decision-maker, should immediately recognise that although she has some numbers that suggest best approaches, there are some issues in the analysis. Some of the data is suspect. The projected sales increases due to more upscale flooring are the result of a series of opinions expressed by the managers, with a fairly crude conversion to profit. She is not sure about the costs of disruption when flooring is replaced. She will recognise that her industry is largely fashion-driven, so replacement is often not a response to the wearing out of materials. It is quite likely that the long-term benefits of the more durable materials may not be realised because fashion changes may dictate replacement before they wear out. She is also concerned about what discount rate to use in the analysis. Numbers can be obtained from the corporate finance department on the WACC for the whole company, but she is not sure that those are the numbers she should be using. Catharine is frustrated and intrigued, and plans to move the analysis to another level.

6.6 Measures Other Than Money

Beyond being just a mechanical tool, discounting processes also help to illuminate decisions.

- Most property decisions involve making expenditure now in the expectation of obtaining some future benefit. A good decision balances costs and benefits – it is almost impossible to arrive at a decision without some balancing process.
- The process helps us to identify what data is important and how much effort should be spent collecting it.
- It tells us to consider time and uncertainty. We inevitably make assumptions about the future, but we know that we cannot know the future with certainty, so some adjustments have to be undertaken.

Discounting techniques are best explained in terms of money. But money merely represents the things that we want to buy with money. We can apply discounting to other measures too, including energy and emissions, and this can be very controversial territory. Various sources point to the role that discounting concepts have had in relation to policy on global warming. Different discount rate frameworks have been used to create a range of possible estimates of the Social Cost of Carbon (SCC), essentially the expected amount of damage caused by each additional tonne of CO_2 released into the atmosphere over time – a number used to help evaluate policy alternatives. Different approaches lead to widely varying values assigned to CO_2 emissions, and hence how much should be spent and how.[1] Lower explicit or implied discount rates imply that more resources should be devoted to energy and emissions matters. A wide range of values has been suggested, including zero or near-zero rates, for example in the Stern Report. Of course the Samuelson railway paradox should always caution against excessively low rates, which would lead society to expend too many resources to achieve uncertain benefits in the distant future and in doing so reduce current consumption, diminishing the quality of life of those now alive.

The elements of discounting should be considered in this context. As environmental issues affect everyone, the private risk factors that dominate entrepreneurial discount rates are largely irrelevant. Considerable uncertainty does remain, in particular surrounding the matter of the possible impact of different levels of emissions, the affluence of future generations as a result of economic and technological growth, the possibility of the destruction of humanity and the emerging capabilities of mitigating damage caused by emissions, or the emergence of other new technologies.

To add to the fog of debate, there are matters of ethical judgement. Evidence suggests that people seem to value those alive now more than those centuries in the future – but is that a correct judgement? How much value should be attached to the preservation of the species? Many practitioners will choose not to engage in this sort of debate, but decisions must still be made – is it alternative A, or alternative B?

6: Notes

1   Jiehan Guo et al. (2006), 'Discounting and the social cost of carbon: a closer look at uncertainty', *Environmental Science and Policy*, 9, pp. 205–16

So We Have
Some Numbers –
But What do
They Mean?

7

Unless you have already decided on a course of action and are simply using a quantified analysis to verify or promote it, some thought should be given as to what the results might mean. Any whole-life costing analysis will help the decision-maker, but further analysis, at least some amount of sensitivity analysis, is almost always in order to fully understand the nuances of what it is telling you.

7.1 Sensitivity Analysis

Any decision-maker, given uncertainty about the economic, environmental and social/cultural returns of various project alternatives, as well as the analytic techniques, might reasonably wonder how different inputs could affect the results.

Sensitivity analysis is the changing of input variables to determine how the results might change as a result. It is an essential add-on tool to basic whole-life costing, as it can illuminate the meaning of initial results. Using it as part of any decision-making analysis is almost inevitable, because most input variables contain uncertainty. It is only reasonable to see what happens if some vary from the expected. It can reveal how different aspects of the multiple uncertainties that typically surround a project might influence the decision, and which factor is of greatest importance. The question could be stated as: 'How far from the "tipping point" is the base assumption? If the discount rate, the life of the solution, or the impact of the emissions were different, would the choice change?' If, for example, the discount rate at which the decision changes is at a number which is highly unlikely, we know that the solution is robust relative to the discount rate.

The more uncertain the inputs, the more important sensitivity analysis becomes. Environmental and social/cultural questions carry considerable uncertainty. What exactly is the ongoing value contributed to society by a park? What is the future impact of emissions? Information about the near future is usually obtainable but, as we all know, even carefully estimated costs of construction can be far off the mark.

The process of sensitivity analysis is like destructive testing. One might imagine testing a complex structure. The proponents might build a piece and load it with sandbags, carefully monitoring it until collapse occurs. A structural designer wants to know how it fails and whether the failure point is anywhere near the range of expected loads to be encountered. If it is, prudence would suggest changing the design – making it a bit stronger in the area that failed, so that one is further away from a professional liability insurance claim. Dealing with a decision-making analysis is similar. By changing some input variables, the 'load' can be increased until the model 'collapses', perhaps revealing that some other alternative is actually more attractive, or that a minor redesign might improve things.

There is copious literature on the subject of sensitivity analysis, and much of it is mathematical and/or theoretical, but dealing with sensitivities in a practical sense can be quite simple. For what might be considered the key uncertainties, including rent, component life, social impact on the local area, CO_2 emissions and energy or capital costs, high and low estimates (best case and worst case) can be made and tested using the model already constructed, and the various outcomes recorded. A set of questions might be addressed: 1) What is the base case alternative that should be selected? 2) What input variables most affect the alternative selected? 3) How does the selection change when the most important input variables are changed? 4) How significant are the differences between the outcomes, given the various changes in the input variables? and 5) What confidence might be assigned to the final selection? A graphical representation can be prepared to support a decision by plotting the output of one or more alternatives against the variable being tested.

Sensitivity analysis, as with destructive testing, can show how systems work. As different input variables are changed, the robustness of the solution can be seen: how far from the base case does a variable have to change before a different alternative should be selected, and how different would the outcomes of that alternative be from the base case?

7.2 Why are Sensitivity Analysis and Other Add-on Techniques Necessary?

Discounting was originally designed to deal with short- and medium-term decisions in which there was little or no managerial discretion. However, building- and land-based problems often have very long-term consequences. Discounting approaches were also not conceived to assist in decisions relating to environmental and social/cultural consequences. Discounting-based methods can exhibit inconsistencies, but there are few other tools that a designer or manager will find as easy to use.

There is ongoing debate about how discounting relates to intergenerational equity. How might well-being be distributed over very long periods of time and shared with people not yet born? How much current consumption should be deferred to benefit them? What might they want? This can lead into lengthy ethical discussions, which may serve more to stupefy than to inform the practitioner.

How mechanistic are decisions anyway? If one builds a large enough computer model, can one take everything into account and find the 'right' answer? This approach assumes that people cannot perform as well as some imagined comprehensive computer model that would have much more memory and processing capacity, and run ideal algorithms. However, it is also possible that a truly advanced computer decision-making tool would mimic the human decision-maker. People appear to have the capacity to imagine more solutions and integrate more functions and subtleties than we could ever include in a piece of software. Possibly the answer lies somewhere between the two extremes, where the ideal is a human decision-maker augmented by computer-based tools.

Computer-based tools do exist and are well worth using, as they prompt the user about what data to collect and make the analysis easier, especially in testing 'what if' scenarios. Research and development initiatives are ongoing, and ever-better software packages can be expected to appear.

What is the 'Right' Decision?

A favourable outcome, such as winning a lottery (or making millions of pounds in a development venture), does not necessarily mean that the 'right' decision was made. For most people reading this book, lotteries will have negative expected outcomes, so participating in them is essentially a waste of resources that could be better used in some other way. While some individual will win any specific lottery, it does not mean participating was the right decision. The probability of success for carefully considered development projects should be higher, but 'winning' does not mean the best decision was made – as with a lottery, it might simply have been the case that a set of unlikely events occurred at the right time.

7.3 Interpreting Input and Output – the Issue of Spurious Accuracy

It is tempting to accept numerical outputs as giving the 'right' answer. In part this is because of the apparent accuracy of any numerical output – spreadsheet software will happily generate figures to as many decimal places as one wishes. Observation of almost any design process reveals something different: an iterative structure, where data is only acquired as necessary, with the initial decisions made very roughly and where highly unfeasible possibilities are eliminated. As the design progresses, more effort is then given to assessing the surviving alternatives. In whole-life costing, when used in decision-making, the same process should be followed – some rough numerical sketches initially, with more detailed analysis as the project unfolds.

Regardless of the method used, it is necessary to interpret the output, which also implies looking critically at the input data – and input data is almost invariably based on uncertain forecasts. Much of Ms Napier's data on flooring consists of managers' best guesses or representations from product salespeople. A practitioner trying to find data on the range of uncertainty about the life expectancy of building products may find no information at all beyond anecdotes and speculations.

Forecasts – which are unavoidable input – always carry significant uncertainty. They tend to be based on trends and discontinuous events, yet 'trend-breakers' are very frequent. They result from such things as economic crises, political shifts, new technologies, new natural resource discoveries, changes in regulation and changes in consumer preferences. In their book de Neufville and Scholtes (2001)[1] point to a need to de-emphasise the importance of developing precise forecasts, and to spend more time creating realistic assessments of ranges of uncertainties and integrating provision for the unexpected.

In the days when computations were done with slide-rules or pocket calculators, people rounded numbers to a workable number of digits, so results appeared as '23,000,000' rather than '22,687,324.27'. Which inspires the most confidence? Given the usual multiple sources of uncertainty about the data input into property decision-making processes, the last six or seven digits should be regarded as meaningless. Of course, it might be useful in some instances to use such numbers to support decisions. People paying for information tend to want accuracy, so there is a temptation for the analyst to comply with a minutely detailed report that can be delivered with a resounding thump on someone's desk.

But People Can be Unhappy with Uncertainties in Estimates ...

One of the authors recalls that some years ago, in creating a preliminary budget, he expressed the per-space cost of an underground car park as a range. The client was incensed about the lack of precise knowledge. At that stage unknowns abounded: soil conditions, site-water conditions, number of parking spaces, how many floors it might have, how it would be accessed, how it might relate to the rest of the building and when it might be constructed. It would have been better to have suggested a specific number (almost any reasonable number), as even if not accurate, it would ultimately have disappeared into the overall project costs, and the controversy could have been avoided.

7.4 Is the Way We Discount Correct?

While discounting-based methods are an attractive approach to using resources suitably, they are subject to criticism, and there are other approaches that are especially relevant to social decisions. Most practitioners will find them intellectually enticing, but of limited use. Ultimately, most techniques still yield a number with which to adjust the impact of expected future events. Even without time preference we have to deal with risk. While we have a reasonable idea of what will happen next week, what will happen ten years from now is less certain, and logically we should ascribe less importance to it.

Discount rates reflect human behaviour, and people are often eccentric. That helps to make design, management and decision-making challenging. If we were all totally rational (including the building decision-makers) things would be much easier (we would agree on things), but less interesting. Conventional economics does not assess why certain preferences or judgements are formed, so complete rationality will be elusive. A change in moral outlook or fashion can influence how rates are formed – perhaps giving different weights to the importance of the welfare of successive generations.

One issue is that much research has verified that conventional 'exponential' discounting is an incomplete approximation of human behaviour, which is especially important in relation to decisions about long-life buildings. Ease of use is one reason why we use conventional exponential discounting; logically it is also consistent over time. However, observation and experiments have repeatedly demonstrated that the discount rate implied by real human behaviour declines the further into the future one looks, following a hyperbolic curve. This is not important when costs and benefits flow over a few years, but becomes of consequence over the longer term, such as when we make decisions about environmental and social/cultural matters with respect to buildings. A number of reasons have been suggested for the existence of this situation. One is that short-term events are easier to visualise than those occurring in the more distant future, so over time we become increasingly indifferent to the passage of time. It is possible to conceive of the needs and expectations of people alive now and those living later in this century – they will be our children and grandchildren. However, it is also understandable that people can become indifferent to whether an event occurs in 200 years' or 500 years' time. Would you differentiate between something occurring in 2113 or 2163?

This implies that the curve by which value falls as described by exponential discounting is incorrect when used over long periods, and a hyperbolic curve is more appropriate (Figure 7.1). As can be noted, over the short decision period of risk-adjusted individual and corporate economic analysis the difference is not significant, but for investments with economic and social/cultural objectives the difference can become significant.

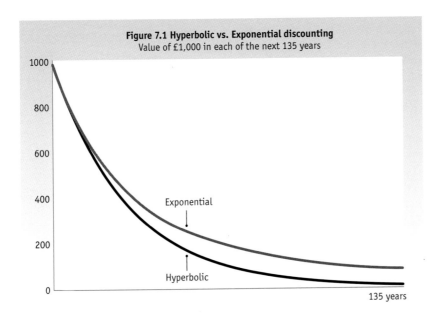

Figure 7.1 Hyperbolic vs. Exponential discounting
Value of £1,000 in each of the next 135 years

For the creators and managers of buildings, consideration of long-term discount rates is important because of the longevity of buildings, especially with regard to sustainability matters. The Treasury *Green Book* recognises this, and suggests the following declining social time preference rates.[1]

- Years 0–30: 3.5 per cent;
- Years 31–75: 3.0 per cent;
- Years 76–125: 2.5 per cent;
- Years 126–200: 2.0 per cent;
- Years 201–300: 1.5 per cent;
- Years 301–: 1.0 per cent.

Of course, using even such low long-term rates still implies that costs and returns in the far-distant future may be given little weight in current decision-making.

This suggests that for events occurring more than thirty years in the future, suitable adjustments should be made, by deducting an appropriate amount from the discount rate being used. Removing 1 per cent from a risk-adjusted discount rate being used for the economic decisions of an entrepreneur will not make much difference, but it might when considering the lower discount rates used for environmental and social/cultural analysis. The practitioner attempting this will recognise why hyperbolic discounting is not more popular: there are logical inconsistencies and the calculations are considerably more difficult.

7.5 Robustness vs. Optimality

In the building world where there is a great deal of uncertainty, there is much to be said for robust design solutions, and this is a key reason to undertake sensitivity analysis. For the designer, it is usually easier to follow a programme based on a projection of specific outcomes, rather than ranges of possibilities. Unfortunately, in most situations achieving a reliable value of the 'expected' outcome, or even measuring the uncertainty surrounding it, is difficult or impossible.

For this reason, a desired feature in most projects is robustness – the insensitivity of the decision to changes in the context of the project – rather than simple optimality, which may depend upon specific conditions unfolding in the future. A search for an alternative that offers the maximum return may imply both limited uncertainty and having the data to obtain precise answers, conditions that are rarely met in the world of land and buildings. Decision-makers should usually seek a robust answer – one that is likely to yield good results given a substantial range of possible future situations. This should be familiar to most building designers: structures are intended to stand even in unusual situations, such as following fire or accidental damage. Architects attempt to ensure robustness in the construction process. Although ideally we would like to design a project and hand it over to a builder and go on holiday, in reality we often incorporate contingency amounts and stay around to solve on-site problems and disputes. A project will seldom exactly follow the planned course when under the assault of reality.

Because of the long life of buildings in particular, and the difficulty in assessing and quantifying the associated uncertainties, in our building world, we are usually seeking robust decisions, not mathematical optimality. Figure 7.2 portrays a situation in which a robust solution might be preferable to the one that offers the best return on the expected numbers, because it will perform well over a wider range.

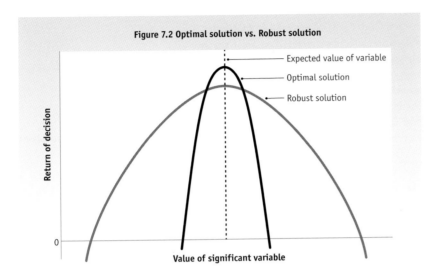

Figure 7.2 Optimal solution vs. Robust solution

7.6 What About the Environmental Issues?

The environmental implications of the creation, use and disposal of products are often explored through life-cycle analysis (LCA). This is a well-developed field, in particular with regard to short-life consumer products for which time and uncertainty are of limited importance. It involves compiling a catalogue of inputs and emissions that relate to the creation, use and disposal of a product, and considering associated implications.

Money-based whole-life costing and environmental LCA are often regarded as distinct methods. They differ in two main ways. Whole-life costing usually does not take into account unpriced 'externalities' – those things such as emissions or other factors not reflected in cash flows. LCA tends to ignore the full implications of labour input, which comprises a large part of construction costs. The units of measure by which LCA operates are not always consistent, with attempts to express environmental consequences of such things as energy, CO_2 emissions, greenhouse gases and the quantity of other undesirable emissions. LCA includes the implications of deconstruction and disposal, while whole-life costing as applied to buildings often leaves these unaddressed.

There are many similarities between the two forms of analysis when applied to buildings. Both deal with human-created artefacts. In buildings there are typically expenditures during the development phase: money for whole-life costing and embodied energy and emissions for LCA. Throughout its life the building will consume more money and energy and generate more emissions. Ultimately, at some unknown time in the future, the building may be disposed of: again, money will be spent and recycling of building components may be undertaken, 'paying back' some of the embodied emissions.

Many of the same problems appear in both forms of analysis – including the uniqueness of each project, dubious and difficult-to-collect data, and long and uncertain life expectancies. Data banks of information exist for both whole-life costing and LCA, but many factors can influence the applicability of available data. In particular, LCA data suffers from a lack of information about the sources of materials and energy for any particular building in a particular place. System boundaries are important in the case of LCA: how far back do you trace the implications of a building, exactly what transport systems were used in moving the materials that were used, and what are the impacts of the labour component? In the case of monetary analysis, all implications are swept into the project through the functioning of the marketplace that priced them. Future uncertainties affect both: when will replacements and refurbishments occur, and what materials and systems might be used then? The assumptions that are made in the analysis, in particular when dealing with events occurring in the future, may affect the ranking of alternatives.

While one might be able to achieve both accurate economic and environmental analyses of mass-produced, short-life, factory-made products, any investigation of building-based alternatives will encompass considerable ambiguity. That uncertainty will increase over time, so in the future actual performance will likely be different from the original projections. The end-of-life situation is highly uncertain, as projections about how, and when, building materials might be recycled or reused some decades or even centuries in the future can be little more than guesses.

A completely sustainable building requires that these disparate factors be pulled together and evaluated against each other. Linking the three factors of sustainability is not easy, but there are ways to work through the issues.

7.7 Have You Got the Right Discount Rate for the Right Phase of the Project?

The insightful individual will realise that the risk is considerably higher during development than after completion. This means that at least two discount rates are necessary – to deal with each phase. The yields at which certain classes of buildings are bought and sold can be obtained from reports from the major estate agents – many reports are available online. This will help to establish a building value at which it can be 'sold' at completion. Just remember that these reflect the market's anticipation of inflation and that each building is different. Some will be new, some old, and some might include development or redevelopment potential.

7 Notes

1   Richard de Neufville and Stefan Scholtes (2011), *Flexibility in Engineering Design*, Cambridge, Mass: MIT Press

All Sustainability Decisions are Complex: Multi-attribute Decision-making

8

The replacement of flooring discussed in Chapter 6 is in many ways a simple problem, but still has numerous complexities. It was only considered on the basis of an economic analysis; most analytical techniques are best explained in terms of money. But money merely represents other things – the things that we want to buy with it: shirts, plumbers, fuels, lunch, architects …

Sustainability adds more dimensions to the art and science of decision-making. The three aspects – economics, environment and social/cultural factors – have to be somehow reconciled in order to avoid either overinvestment or underinvestment. As usual, there are trade-offs to be made. A building design or system incurring high emissions during initial construction but low emissions during its service life and disposal may have to be compared with one with lower emissions in construction but higher emissions in use. A durable design may perform better over the longer term, despite higher emissions at the start. This is not always the case, though – investing more doesn't always give good value.

One common approach is to monetise the unpriced attributes of available alternatives, but many object to this. There are numerous issues surrounding both monetary and non-monetary approaches. Personal belief is important: does the decision-maker believe in consumer sovereignty, or that some other evaluative process can arrive at a superior valuation of assets and services? Probably the answer is somewhere in the middle, and the extremes are to be avoided. Regardless of this debate, there are established ways of dealing with multi-attribute problems.

Land and buildings can be seen as a 'bundle of attributes' – and quite a complex bundle, at that. Sustainability, meaning economic, environmental and social/cultural issues are always interwoven – every building decision will involve all three, and in virtually all cases there will be a conflict between attributes. Some alternatives will score high on an economic scale, while others may outperform them on environmental or social/cultural scales. But all factors must perform to some level. A building that is designed to be very 'green' but fails to perform economically may be demolished or dramatically rebuilt within only a few years. Similarly, a building that fails on a social/cultural basis may become a haven for crime and disorder, or people may simply come to regard it as ugly, again leading to premature replacement or substantial reconstruction. Hence, even a 'green' agenda demands performance in the other dimensions.

This issue can be found everywhere. One only has to look at ranking systems – for example those that rate cars, universities or cities. In order to understand this, it is necessary to put issues of uncertainty to one side, and consider some techniques for making decisions under conditions of certainty.

8.1 Where Do You Want to Live? Vienna or London? Düsseldorf or Baghdad?

Numerous quality-of-life surveys are produced to assist people and companies in choosing whether to locate in one city or another. Presumably a successful city will rank near the top, but any person using such surveys might wonder how ranking systems actually work.

According to various ranking systems, the following cities tend to rank highly: Adelaide, Auckland, Calgary, Copenhagen, Geneva, Helsinki, Melbourne, Munich, Perth, Sydney, Toronto, Vancouver, Vienna and Zurich. These might be seen to be 'sustainable' cities – they are prosperous, they offer opportunity and their inhabitants have exceptional lifestyles; they also have decent environments and offer other social/cultural benefits. Baghdad, Harare, Khartoum, Kinshasa and Port-au-Prince feature near the bottom of most lists. They miss on almost any measure, including environmental scales.

But the rankings differ, because they are highly dependent upon the weighting of various criteria. *The Economist* uses thirty factors in five categories to create its rankings, but other systems use different criteria and assign different weightings. Some top ten lists are dominated by cities in Australia, Canada and New Zealand, while European cities dominate others. Would one of these systems be of value in considering where you might like your company to send you? Perhaps, but most people would also give regard to their personal priorities. Suppose you especially like music, or surfing, access to wilderness, mountain climbing or skiing. Imagine you were very worried about personal safety – Luxembourg ranks very highly in that respect. There are also systems that rank cities from a student's perspective; and here London and Paris receive high scores. One may put an absolute filter on the results, perhaps requiring that the city to which you might move uses a language you speak.

Making decisions about buildings is like ranking cities. Numerous alternatives need to be dealt with, and this involves identification, initial screening, prioritisation, analysis, ranking and selection. Many attributes may have to be identified. In almost any such 'multiple attribute' problem there are different units of measurement and, in the design world, objective quantification of some attributes may be impossible or impractical. If you do not want to simply follow someone else's predilections, how do you select a city, a design or a building component?

8.2 Complexity is Nothing New in Decision-making

Difficulty in making such decisions is not a new experience. Polymath Benjamin Franklin (1705–90), when responding to a request for advice from Joseph Priestley (1733–1804), the clergyman, educator, natural philosopher and discoverer of oxygen, wrote a letter – not suggesting an answer to the problem at hand, but outlining how to solve it. Franklin saw that one problem in making decisions without some system was that 'all the reasons pro and con are not present to the mind at the same time; but sometimes one set present themselves, and at other times another, the first being out of sight'. Franklin offered a set of steps:

1. Identify the question.
2. List the pros and cons of the possible alternatives.
3. Assess the importance of the pros and cons.
4. Eliminate the unimportant attributes so that the more significant differences become more apparent.

Franklin implied a number of other aspects of the process, including:
- the need to assess the probabilities associated with the pros and cons. Some of the pros and cons are uncertain and may not occur. The ranking of uncertainty can occur, perhaps on a scale from one to ten, with five being fifty–fifty;
- assignment of weightings to the pros and cons, perhaps with 10 being assigned to the most important and 1 to the least;
- reflection as an integral element in the process. Franklin suggested that 'consideration' be given over a period of days.

He saw an important tool as being the process of creating a table of the various decision-making elements, as it requires one to reflect on the issues. Questions that might be asked about the process, data and weightings include:

- What patterns emerge?
- What factors really matter? Have we accounted for all of the factors?
- What factors absolutely have to be fulfilled?
- How important is the decision? What further action does it warrant? What additional information might be worth acquiring?
- How much time should we invest in getting a better answer?
- What are the risks, and what is the willingness to take them?
- Does this decision relate to any other decisions? If so, how?
- How might the assessment differ for others who are influenced by the decision, such as the organisation, the client, the owner or the users?

8.2 Building on the Basic Model

The designer or manager will probably follow one or more decision-making rules, whether explicitly or tacitly. Different rules fit different contexts, based on availability of data, cost of obtaining more data, quality of data, personal belief and probably the perceived significance and consequences of the decision being faced.

Ranking some decision-making rules in order from the crudest (most easy to use) to the most sophisticated, they might include the following.

1. **Simple recognition:** This is undoubtedly the most common decision method. We may pick out a bottle of jam because we bought it before or recognise the brand name. This method can work well when the decision is not of great consequence, and/or the decision-maker is risk averse, or when gathering data to use in a more complex analysis will be difficult or expensive.

2. **Satisficing:** This (the word combines 'satisfy' with 'suffice') is used to sort alternative possibilities into 'acceptable' and 'unacceptable'. It does not seek an optimal solution, but by identifying alternatives that are 'good enough' attempts to reduce the number to a manageable level. Most tests are satisficing: on university courses, while it is necessary to pass all the courses to graduate, it is not necessary to achieve the highest overall mark. Similarly, most buildings are acceptable as long as they meet the various expectations of the people associated with them – presumably being more-or-less sustainable according to the three categories (economic, environmental and social/cultural), although different people may weight these differently. It is usually not necessary for a building design to be 'the best' to be acceptable. Often achieving 'the best' (whatever that may be) is just not worth the time or effort.

   a. **Conjunctive rule:** This variant of satisficing involves setting cut-off points on the important attributes and reviewing the alternatives with respect to these points. This process eliminates alternatives that do not 'pass' every important attribute. The selection of cut-off points can be important. Too high and no alternatives may pass; too low, and few alternatives may be eliminated.

This rule can be used in a very simple manner for less consequential decisions. One decision-making rule might be 'to select the first alternative that meets all the minimum cut-offs'. Any unanalysed alternatives would remain unconsidered, even if they were superior. To the consultant this could be useful, given that scores and weightings might be based on guesses about the preferences of others – especially the client – so may be only approximate.

b. **Disjunctive rule:** In this case an alternative is chosen only if it exceeds the cut-off value on one or more attributes. The notion is that the alternative is required to perform well on at least one attribute, overall performance is not demanded and 'failures' will be tolerated. Of course, the issue for the property decision-maker is that unless the cut-off values are set very high, a substantial proportion of alternatives will pass the test. If they were not seen to be likely to perform on at least one attribute they would not likely have been included in the decision set.

3. **Lexicographic methods:** These techniques gained the curious name 'lexicographic' because the process is similar to the way in which words are located in a dictionary. To find a word one looks for the first letter, then the second, then the third … In a lexicographic decision-making method one first considers the most important aspects of the decision, presumably those given the highest 'weightings'. There are different ways of implementing such a method.

a. **Selection by aspects/Take-the-best:** The decision-maker selects what appears to be the most important attribute, and selects the best-performing alternative with respect to this. If several tie, or perform well with respect to the attribute, a second important attribute is selected, and the best alternatives are selected from the reduced decision set. At some point, the 'winner' will become apparent, perhaps because only one alternative is left.

b. **Elimination by aspects:** When this approach is employed, one picks an attribute that appears to be important, and determines a 'cut-off' point for it. All alternatives that fall below that point are eliminated. Then what appears to be the next important attribute is selected, and alternatives below that cut-off are thrown out. This process is repeated until the decision becomes apparent.

4. **Additive weighting method (multi-attribute utility theory):** Most decisions in the property world involve selecting between alternatives: carpet or vinyl, tile or paint, developing an office building or a hotel … The Franklin method can be readily extended to decide between any number of alternatives. Given the reality of opportunity costs, even go/no-go decisions imply the selection of something on the no-go side, even if it is just leaving money in the bank and the site vacant. Additive weighting methods consider a set of what have been determined to be important attributes of the various alternatives, but do not necessarily assign them equal weight. Unlike the simpler methods,

this approach is 'compensatory', in that a good performance in one area can compensate for a poor performance in another.

The first step is to determine what attributes are important – perhaps goals that might be met by the project. They could be classified according to the three categories of sustainability, and perhaps a category for concerns that do not fit neatly into any of the three (Figure 8.1).

Figure 8.1 Goal hierarchy for sustainable building component evaluation (Sample factors)

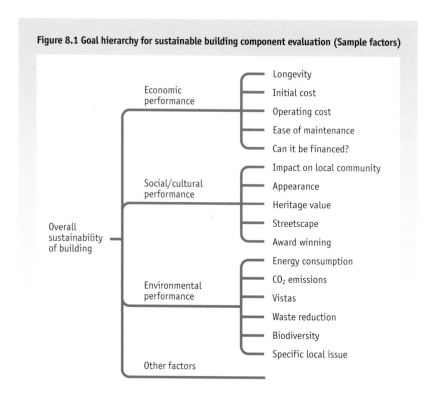

From Figure 8.1 a table can be created to compare alternatives (Figure 8.2). The decision-maker sets a weighting, 'W', for each factor, perhaps on a one to ten scale. For each attribute as it relates to each alternative, a score, 'S', is created, on the basis of quantified data or qualitative information – perhaps gut feel. These are reflected upon, multiplied together and summed up. The results give some insight into how the alternatives might be ranked, with the highest number presumably ranking first. This is essentially how the ratings of cities is undertaken, albeit using different categories.

This reflects a fundamental problem in all such analyses, including those embracing issues of sustainability. Mathematical analyses demand some sort of common measure of the different factors. One to ten relative weights and scores are dimensionless, but a great deal of subjectivity can appear in both the weightings and the scores (how exactly does one easily quantify different views on the importance of waste reduction, for example?).

Real quantitative data is rarely expressed on a one to ten basis – it tends to be counts of money, or energy, or square feet, or some other performance measure. In some of these cases, data can be normalised into dimensionless units, and again the higher numbers express some notion of being 'better'.

Figure 8.2 A sample multivariate decision matrix (Sample factors)									
Which wall system?		**Alternative A** Low initial cost			**Alternative B** Expensive/long life			**Alternative C** Green		
Factor	**Weighting (W)**	**S**	**W**	**I**	**S**	**W**	**I**	**S**	**W**	**I**
Economics										
Longevity	7	6	7	42	8	7	56	4	7	28
Initial cost	9	8	9	72	4	9	36	4	9	36
Operating cost	7	6	7	42	8	7	56	5	7	35
Ease of maintenance	6	5	6	30	9	6	54	3	6	18
Can it be financed?	8	7	8	56	8	8	64	7	8	56
Social/cultural										
Impact on local community	5	7	5	35	8	5	40	8	5	40
Appearance	5	4	5	20	10	5	50	2	5	10
Heritage value	2	8	2	16	6	2	12	2	2	4
Streetscape	8	5	8	40	5	8	40	5	8	40
Award winning	4	4	4	16	8	4	32	8	4	32
Environmental										
Energy consumption	7	4	7	28	6	7	42	9	7	63
CO_2 emissions	6	4	6	24	5	6	30	8	6	48
Vistas	1	6	1	6	5	1	5	4	1	4
Waste reduction	5	5	5	25	3	5	15	8	5	40
Biodiversity	2	6	2	12	5	2	10	6	2	12
Specific local issue	5	4	5	20	4	5	20	8	5	40
Other factors										
Total weighted score:				**484**			**562**			**506**

The high score is preferred. Factors from one to ten, with ten being seen as the 'best' in each case.

Where: S = Score; W = Weighting – relative importance of that factor to the decision; I = S x W

This method, as with all the methods, has strengths and weaknesses – which means that, again, it cannot be relied upon to give the 'right' answer.

- It does not take into account interactions between the attributes; it assumes they are independent. The results of a survey might show that people prefer brick exteriors over metal, and traditional design over modern design; however, the two factors may interact. Perhaps they only prefer brick when the exterior is a traditional design and may prefer metal for a modern design. It is often difficult to understand interactions between attributes, but fortunately in most cases they are not of great consequence – unlike interactions in the drug industry. Nevertheless, the possibility of interaction needs to be kept in mind, especially when making decisions of consequence.
- Creation of the scores and weightings can be difficult. While sometimes real numerical data can be obtained, perhaps price per square foot, annual maintenance costs, projected energy costs or CO_2 emissions, others will inevitably be the result of subjective evaluation and, as was seen earlier, data collection can be subject to confirmation bias.
- Many (or perhaps most) scores and weightings will be based on assumptions and subjectivity.

The strength of this method is that it forces the decision-maker to identify and separate the various aspects of the decision, put some figures on them, and then do a mathematical analysis. Moreover, it can be done at any level – one typically starts with a rough chart and then refines it if necessary. And of course, sensitivity analysis (see page 69) can help to define a robust answer.

5. **Dominance:** This is an important concept, and it helps to explain why there are so many products available and apparently viable in the marketplace. Attributes often compete with each other, and a product that is 'green' may require more maintenance or may not last as long as something else. Few products dominate all others in all important areas. Hence, the product selection will depend upon the individual doing the selecting and on his/her priorities. Domination clearly occurs if one alternative surpasses another on one or more significant attributes and at least equals it on all others. Therefore one approach is to compare alternatives two at a time. In such a pairwise analysis, if one alternative is dominated by another it should be eliminated from consideration. Then the remaining alternative is compared with another, and so on until a non-dominated set of alternatives is obtained. This is rather like a tennis or squash ladder, where two people play at a time and the winner moves on to the next competition. While any non-dominated alternative can be a rational choice, the reality is that one particular alternative may be dominated by another, while actually being better than a different non-dominated alternative – which is why it is best to use this approach alongside one of the other methods.

6. **Other methods:** Beyond all these listed are other methods, but they tend to be the subject of specialists and agencies making very large decisions, so have corresponding budgets for data collection and analysis. Most architects, developers and builders do not need to go into them, although some have their attractions, including appealing graphical output.

Such methods are part of the portfolio of tools that we use in approaching decisions. A practitioner may never actually map out decisions on paper or use a computer, but understanding the concepts will help move him/her to a higher capability. There are some potential hazards to be pointed out, though. Although one might initially think that adding information would yield a better decision, this is not always the case. Some research has shown that good decisions may be made with knowledge of only a few important properties. Adding information about more attributes may include some that are irrelevant, and may confuse the decision-making process. Comparing four or five different product alternatives on the basis of four or five properties is not necessarily an easy task, and is more than most people can cope with, mentally – which is why some of the simpler selection methods are so valuable.

Although our teachers and parents warned about the dangers of procrastination, there may be valid reasons why this 'bad' human habit is so persistent. Often good decisions are made after having 'slept on it'. Franklin suggested more than one period of reflection during the process in his letter to Priestley. Time, distractions and sleep seem to be good incubators of ideas and solutions, although the mechanisms by which they work remain obscure.

8.3 Shadow Pricing: Relating Monetary and CO$_2$ Emission Measures

One technique for linking different measures is 'shadow pricing' whereby one factor is assigned a value in terms of the other (usually in monetary terms). The 'Shadow Price of Carbon' (SPC) attempts to capture the amount that society might reasonably pay to reduce emissions. Different studies, using different methods, different assumptions about future costs of climate change and how to relate them to the present, have led to widely varying monetary values being assigned to emissions in an attempt to put a price on their impact.

The SPC set by DEFRA (UK Department for Environment, Food and Rural Affairs) in 2007 was £25/t CO$_2$ e (equivalent carbon dioxide), rising at 2 per cent per year. Other projections range from £3 to £150.[1] The idea is that shadow pricing allows at least some of the unpriced implications of a project alternative to be brought into an overall project analysis. Of course others, including a wide range of factors such as species protection and easing urban congestion, will remain outside the analysis.

While discounting is the normal way of dealing with money-based decisions, it is not well established for dealing with environmental matters – even those we can count, such as CO$_2$ emissions and energy consumption. There is considerable debate surrounding how future CO_2 emissions might be discounted in order to create a shadow price. One extreme view is that environmental matters should not be discounted: that CO_2 emissions one hundred years from now should count for as much as those of this year. Of course, the Samuelson railway example warns that if there is no adjustment for time and risk, society would spend all of its resources in order to gain benefits centuries in the future and reduce current consumption to zero. Logically, the discount rate needs to be positive.

Ongoing debate should be expected with respect to how to approach the time and risk issues associated with long-lived assets. However, again, the designer does not have the luxury of waiting for the academic world to produce a definitive answer – decisions have to be made now about which alternative to select, and the use of shadow pricing and discounting is accessible to the practitioner, even if the numbers lack certainty.

8.4 Beware of Slapping Lipstick on a Pig

So far, concepts and tools have been presented that should be of assistance to the individual attempting to make sustainable decisions about property. But it should also be obvious that there is rarely a completely 'right' answer. Multiple attributes, uncertainty and time issues make the entire process complex and ambiguous, so ultimately many decisions are at least partially made on the basis of individual judgement – hopefully experienced, expert and insightful, and supported by one or more of the available tools.

It is the multivariate nature of the decisions that makes sustainable design so difficult. At every step there is the problem of reconciling the relative importance of economic, environmental and social/cultural issues.

One hazard along the way to achieving true sustainability is that it is tempting to make the trade-off in ways that support specific agendas. If one is selling a 'green' product, for instance, or is a politician attempting to be re-elected, then some manipulation may be used – perhaps unconsciously – to tip the results. In particular, it can cause the over-valuation of what might be described as 'fuzzy hopes'.

Quantification, when combined with sensitivity testing, is one way of addressing this problem. Although many things are difficult to quantify, attempts to attach numbers to them can reveal major discrepancies. Thinking back to opportunity costs and limited resources, choosing one thing almost always implies not choosing something else – and one does want to end up with the best package at the end. We may not always like it, but money does provide a convenient scale with which to compare alternatives. Many goods and services are explicitly priced in the marketplace, and the history of governments attempting to set values is very discouraging. Even if there is no explicit price for something, there may be an indicator about its worth, and we might attempt to define reasonable estimates of feasible ranges for others. Sensitivity testing can help indicate whether the ranges given for those difficult-to-quantify factors are important in determining the results, including how close the decisions might be, and why, and whether more data should be sought.

It is vital to question how risk trade-offs might be made. Is it better to take an alternative offering a lower expected outcome over one with a higher, but less certain, outcome? The answer has to lie with the person or organisation making the decision, but asking it is always necessary.

8: Notes

1   Jiehan Guo et al. (2006), 'Discounting and the social cost of carbon: a closer
     look at uncertainty', *Environmental Science and Polity*, 9, pp. 205–16

A Simple Building Example

9

The practitioner might consider two ways of relating analyses of economic, environmental and possibly social/cultural matters to make an overall evaluation of the sustainability of project alternatives. One is to assemble the various flows of resources associated with the alternatives under consideration, and compare them graphically, keeping in mind the indifference curves considered in Chapter 4. The other is to use shadow pricing.

9.1 Alternative A or Alternative B – Exploring Multiple Criteria

A client has approached an architect to design a simple regional distribution warehouse. It is to measure 5,000 square metres, of which 100 square metres will be office space. Three different designs have been created: there is the basic warehouse possibility (alternative 1), but for more construction cost, CO_2 emissions can be reduced, both initially (the embodied CO_2), and on an ongoing basis due to reduced consumption of electricity and gas. At this point, it might be decided to put the demolition and disposal costs to one side – it is a simple building expected to last sixty years or more, perhaps longer in some other use.

Some expected flows of money and CO_2 can be assembled for the basic building, for its lifetime (Figure 9.1).

Figure 9.1 A simple warehouse with associated office area

5,000 Sq m including 100 Sq m office area

A: Construction — Alternative 1

Element	Economic cost £	% of cost	CO_2 'cost' embedded CO_2 kg	% of cost
Substructure	339,250	16.9%	735,000	46.4%
Frame & upper floor	359,900	17.9%	340,000	21.5%
Roof	294,000	14.7%	210,000	13.2%
External walls	88,500	4.4%	65,000	4.1%
Internal walls	43,750	2.2%	10,000	0.6%
Finishes	22,000	1.1%	30,000	1.9%
Services	184,000	9.2%	150,000	9.5%
External works	465,750	23.2%	45,000	2.8%
Other	209,000	10.4%		
Total	2,006,150		1,585,000	
	401.23/sq m			

B: Ongoing operations

Building-related costs only (air-conditioning only in office area)

		Economic cost	CO_2 'cost'
Gas (GJ)	1,473	11,457	84,288kg
Electricity (kWh)	146,059	13,145	82,085kg
		24,602/year	166,373/year

C: Replacements

	Economic cost	CO_2 'cost'
Roof (year 25)	294,000	105,000kg
Parking surfacing (year 20)	80,000	40,000kg

The observant individual might notice that the building components of greatest consequence in each area are different. Substructure represents 46 per cent of the CO_2 embedded in the project, presumably a result of the concrete used in the foundations, while only representing 17 per cent of the cost. This is important, because it suggests how one might modify the design. The designer could focus on reducing the CO_2 consumed by the foundations; perhaps the overall cost implications would be minor (foundations are a minor cost element), but he/she could compensate for it by finding some capital cost savings in external works – the highest cost area, but that which is associated with the least CO_2 consequences. It is always easiest to find reductions in areas which account for bigger portions of the overall budget, no matter whether an economic or environmental improvement is sought.

For a combined analysis, two more designs have also been developed: both have lower CO_2 emissions and operating costs but, not unexpectedly, cost more to build (Figure 9.2). Which one should be selected?

Figure 9.2 Summary of alternatives			
Percent different than Alternative 1			
		Alternative	
Cost implications:	1	2	3
Economic			
Capital	100%	105%	115%
Operating	100%	90%	85%
Environmental			
Capital	100%	100%	100%
Operating	100%	90%	85%

Environmental and economic costs are incurred both in the short-term future, and in the longer term, suggesting that time and risk have to be dealt with. The two-sided nature of this analysis becomes apparent, and discounting the economic costs at a higher level than the environmental implications is in order. In this case real rates of 9 and 3.5 per cent respectively were used, with a sixty-year study period.

A number of mathematical probes can be made into this question, including preparing discounted analyses of the economic and environmental performances. One might reasonably start with the social time preference rate for the environmental decision, and a real risk-adjusted rate relating to the economic aspects, perhaps 3.5 per cent and 9 per cent respectively.

The results of one analysis can be plotted (Figure 9.3) on axes of economic performance and environmental cost. (Note that the graph shows costs as negative numbers. This means that the best situations are in the upper right-hand corner. Construction practitioners, used to thinking in terms of costs, might choose to reverse the graph.) Of course, it would be great not to forget the third dimension – social/cultural matters – but given the limitations of two-dimensional paper, here we will just work with two factors. This alone does not give an answer; we have to know how the two factors are traded off, and that can be quite personal, or corporate. The developer will be hesitant to sacrifice economic performance (more costs) unless he can expect a considerable increase in environmental performance – line X–X. An environmentalist would put more relative value on environmental performance so would make trade-offs on a line with a different slope (line Z–Z). The embattled architect might attempt to compromise on a line with an intermediate slope (line Y–Y). What the solution might show, is that some design solutions (such as some alternative D) may be clearly dominated by others, no matter which trade-off line is chosen.

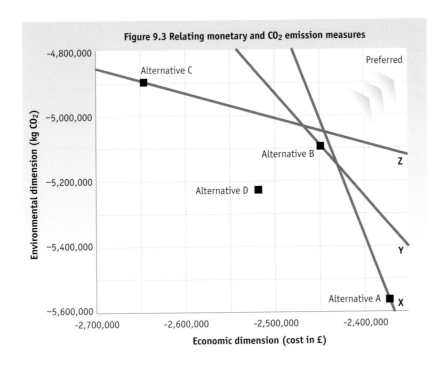

Figure 9.3 Relating monetary and CO₂ emission measures

9.1.1 Shadow pricing

Shadow pricing is essentially the same thing, done a bit differently. The cost of the CO_2 (or CO_2 equivalent) emissions can be added to the project's economic analysis, which can then be inspected to see what the implications might be. The shadow price of the CO_2 is analogous to the slope of the line trading off economic performance against that part of environmental performance. There is considerable debate about what the 'right' number might be. An individual focusing on economic performance would prefer a lower number; an environmentally oriented person would choose a higher one. A sensitivity testing around the question would show what the shadow price needs to be in order to change the decision. An analysis of the warehouse problem using various shadow prices indicates that the decision should be to build the basic alternative until the shadow price of carbon was somewhere over £600 per tonne – higher than even the most extreme estimates – suggesting that building alternative A would be a reasonable selection.

9.1.2 Is this all there is?

Of course not: the concepts, factors, data and layers of analysis that can be added seem almost infinite. There has been much academic work addressing small details of this process. Environmental concerns encompass much more than CO_2 emissions, and the analysis described does not embrace them. The issue is that it is exceedingly difficult to subject many environmental concerns to quantified analysis – energy and CO_2 are good subjects for analysis because they can actually be counted, although there is considerable uncertainty about how to measure the energy and CO_2 embodied in buildings.

Of some concern is the potential for double counting of environmental and social factors. This is because some of their implications may already be included in an economic analysis, often through a tax or other charge. For example, builders pay disposal fees, so some economic compensation is being made for construction waste generation; utility costs include some element to pay for emission reducing technologies; and materials costs may embed some element to pay for remediation of the disruption caused by mining.

Given multiple uncertainties, a Monte Carlo simulation could be run to explore the different possibilities that might unfold in the future. These might include such matters as future energy sources and costs, project or component life expectancies, the system boundaries for the environmental analysis and the users' value systems, including both their assessments of time preference and risk aversion (which affect the discount rate chosen) and how they weigh economic and environmental matters.

The formulation of the analysis requires some forecasts about what is likely to happen during the study period, but forecasts always carry significant uncertainty, and this uncertainty increases the further one looks into the future. While discounting of costs is in many ways crude, it does help us in focusing on the shorter-term events, about which we might be able to get good information.

Monte Carlo Simulation

Monte Carlo simulation was developed to deal with problems that contain significant uncertainty with respect to inputs, and when a clear formula does not exist that will give the 'right' answer. Even though there is considerable theory and literature surrounding this practice area, it is quite accessible to practitioners, who can create simple simulations using a spreadsheet program. In the initial stages of solving any problem, simpler is better – what you often need is illumination about the nature of the decision, and building a basic model that encompasses the most important aspects of the problem can be invaluable in this respect.

Essentially, Monte Carlo simulation involves building a numerical model of the process about which one is trying to make a decision – this is the simulation. In a typical problem there are one or more uncertainties which affect the results; in the warehouse problem, for instance, these might be the price of electricity and gas, and the longevity of the building. While we cannot know what they will be, we might make some assumptions about how they move based on historic data or personal experience. One uses the spreadsheet's random number generator to draw randomly from that distribution and those numbers are used in the model to create one possible scenario. By doing this numerous times a distribution of results appears.

This process does not tell you the 'right' answer – it is rarely that obvious, even with hindsight – but it does illuminate possible outcomes. Based on individual or corporate levels of risk aversion one might make a selection, or decide to collect more data.

Pulling the Decision Apart: Four Simple Assessment Matrices

10

Earlier sections looked at sustainability in terms of its three components – economic, environmental and social/cultural – and at how some decision-making techniques might be used to help resolve design quandaries and arrive at more sustainable solutions. How to deal with significant issues such as uncertainty and poor, difficult-to-collect data has been addressed.

Yet, we know that most decisions are undertaken without numerical analysis. A designer may be making hundreds of small decisions a day, so complete quantification of each is not feasible. Often we do things simply by doing what we did before; if that roof specification worked before it will probably work again, so we do not have to revisit the question every time. Ms Napier, in analysing her flooring choice, reasonably expects that if she can come up with the 'right' answer the company will use it repeatedly, without having to revisit the question every time they replace flooring. That justifies her effort.

What is surprising, though, is how often, when making a significant decision, we do not even write down the pros and cons as suggested by Franklin. How does one make a significant purchase – perhaps a car? It is possible that the quality of the decision might be improved by writing down a few things, and applying a few rough numbers, perhaps helping you avoid some of the decision traps previously discussed.

Disaggregation, or breaking down a decision, can help in visualising the component parts and putting them into perspective. But where should one stop? One of the biggest questions is how much analysis and data collection should be undertaken.

A simple tool is used in order to help you decide how much effort to put into a decision and to understand problems. Accordingly, a set of assessment matrices can be used as a framework within which to assess whether or not to invest effort into the decision, by compelling the decision-maker to focus on the costs, benefits and possibilities related to any specific decision. It is designed to work with personal gut feel, by forcing simple consideration of the individual elements. Quantification is avoided.

10.1 What is the Potential for Sustainability in the Project?

To be sustainable, a design decision must produce a benefit in terms of at least one of the three aspects of sustainability.

There may be benefits in one kind of sustainability but costs in another. The trade-off between costs and benefits occurring in different kinds of sustainability can be difficult – but is often necessary. If there is limited potential to increase the sustainability of the project, it is probably not worth spending the money or effort.

An approximate notion of sustainability potential can be obtained by scoring this. Put an 'X' in one of the five boxes for each type of sustainability, and after that put an 'X' in the lower box to give some notion of the overall sustainability potential of the project.

Figure 10.1 Sustainability potential

	Large gain	Some gain	Nil or gain	Some loss	Large loss
Environmental sustainability					
Economic sustainability					
Social/cultural sustainability					

	Large gain	Some gain	Nil or gain	Some loss	Large loss
Sustainability potential					

10.2 Motivation: the Investor

Architects and other consultants advise clients and implement their demands and inclinations, but it is the clients that make the primary investment decisions. In the necessary dialogue it is important that the architect understands the client's point of view.

Attitudes towards sustainability vary widely, and can be expected to change from time to time. An individual or organisation that thinks action on sustainability is urgent is more likely to put resources into sustainable investment than one that thinks action is not urgent. Investors who take a far-sighted view of the performance of a project are also more likely to put resources into sustainable investment. A decisive investor is more likely to commit resources to a preferred course of action, even though there may be risk or uncertainty attached to the commitment. Resources are required for any investment, so a well-resourced individual or organisation is in a better position to undertake sustainable investment than one that is poorly resourced.

Figure 10.2 The investor

	Urgent		Neutral		Not urgent
Urgency of action on sustainability					
	Far-sighted		Neutral		Short-sighted
Far-sighted or short-sighted					
	Decisive		Neutral		Cautious
Decisive or cautious					
	Well resourced		Neutral		Poorly resourced
Well resourced or poorly resourced					

	High		Neutral		Low
Motivation for sustainable investment					

10.3 Feasibility: the Building (or Part Building)

Extra investment for sustainability should only be made in designs or decisions where there is a good expectation that it will be repaid by the benefits received through use.

Long-life buildings and building elements give more scope for initial investment in sustainable features to pay off during the operational period, compared to short-life buildings. Of course, different elements in buildings have different life expectancies. Shop interiors tend to have short lives driven by tenant turnover and the dictates of fashion. Mechanical systems last longer, and the frame of a building longer still.

Stability during the building's operation provides a better opportunity for the benefits of sustainable investment to be achieved. The assessment matrix identifies three important factors that might disrupt stability. For all three factors, stability favours sustainable investment.

As was seen in Chapter 3, sustainable investment is more easily justified if the operational benefits can be set against the investment cost, and harder to justify if construction and operation are separated.

Again, your individual assessment of the project is made by marking an 'X' in the appropriate boxes. This will yield your overall assessment of the project opportunities for sustainable investment – broadly, a reflection of the upper boxes.

Figure 10.3 The building

Long life or short life	Long life		Neutral		Short life
Stability of use	Stable		Neutral		Volatile
Stability of regulations	Stable		Neutral		Volatile
Stability of technology	Stable		Neutral		Volatile
Linkage between construction and operation	Integrated		Neutral		Separated
Motivation for sustainable investment	Strong		Neutral		Weak

10.4 Combined Assessment

The overall assessment of whether a construction project has a realistic chance of making a positive contribution to sustainability depends on all of the three factors that have been reviewed individually: the sustainability potential, the investor and the opportunity offered by the building itself.

Figure 10.4 Combined assessment

	Large benefit	Some benefit	Nil or equal	Some cost	Large cost
Sustainability potential					
	High		Neutral		Low
Motivation for sustainable investment					
	Strong		Neutral		Weak
Feasibility of sustainable investment					
	Strong		Neutral		Weak
Overall project sustainability evaluation					

10.5 Where to From Here?

Is this excessively simple? Yes, but that is the point. A non-quantified approach can help to inform the decision-maker about the nature of the decision and the appropriate level of effort to put into further analysis. There are many obvious next steps. One would be to attach numbers to the process, with weightings of relative importance, using subjective opinion, much as Franklin proposed. More factors might be included. A designer/decision-maker might like to explore his/her own attitudes, perhaps to see how they accord with the overall evaluation and with the client's.

Learning from the Past: Heritage and Sustainability

11

11.1 Quandaries

The first quandary is that while the tradition has been to discuss creating new buildings, architects are increasingly dealing with existing buildings. Buildings are one of the few assets that usually continue in 'normal' use for decades, or centuries, so in any discussion of sustainability, it is necessary to give some consideration to what happens to them over time. Only a small amount of the building stock is added each year: in 2011 in England, with a stock of 22.8 million houses, 117,700 new-build dwellings were added – about half a per cent. When population and economic growth are low the demand for new buildings decreases. Hence, one can, in many places around the world, find communities which have stood still – sometimes for centuries – and have retained their old building stock. In advanced societies, special value has come to be placed on older buildings.

The second quandary is that as a result of their long lives, most buildings will not incorporate the latest materials, details and equipment, so may be less than ideal from an energy-efficiency perspective. Older buildings were usually built single glazed and with minimal insulation. This is not to be deplored because their creators were usually behaving rationally given the situation in which they were building – to have spent more in days of coal heating, few insulating products and limited window choice would probably have constituted overinvestment.

One approach is to demolish everything and replace it with new construction. While possibly attractive as government policy as a way to reduce energy consumption and stimulate the building industry, it has not become accepted, and has a number of shortcomings. Relative to sustainability, this one-size-fits-all approach focuses almost entirely on environment and evades economic and social/cultural issues. The cost of replacing a nation's entire building stock would be astronomical, and would imply passing up other opportunities, some of which might yield higher energy savings – perhaps upgrading public transit networks, for example. In the larger view of sustainability, it might take away investment from such worthy alternatives as upgrading the educational system. Devoting large amounts to building replacement to achieve energy savings might constitute overinvestment relative to the alternative possible uses of those resources.

11.2 Evaluating Heritage Value

Coping with heritage buildings and urban environments forces one to confront the wider nature of sustainability, and not just focus on environmental matters. The *Guide to the principles of the conservation of historic buildings*, BS 7913:1998, embraces the three areas of sustainability by seeing objectives in economic, environmental and social/cultural terms.[1] The complexity of heritage is that much of the return is in the social/cultural area, the element that is most difficult to quantify.

The existing building stock makes a significant contribution to social/cultural well-being. Pearce included heritage buildings (and other elements of built environments) as part of the human-made category of assets.[2] In keeping with the multitude of attributes that buildings encompass, he saw them as not just augmenting economic processes (much as machine tools might), but as making contributions to human well-being in their own right, much as reading, learning and education can be enjoyed, and are not just contributors to increased production of other goods and services.

As a society we make many investments in the expectation of some sort of social return – that is happening whenever a library or museum is created. One presumes that a country is better for having museums, art galleries and libraries. People, expressing themselves collectively through political and other processes, demand cultural assets.

Such benefits are difficult to quantify using conventional tools. The English Heritage guide, *Conservation Principles, Policy and Guidance*, explores and classifies the values to be found in heritage places. They note that sometimes there is a conflict between 'utility' and heritage value, that 'Utility and market values, and instrumental benefits are different from heritage values in nature and effect.'[3] The document proposes that value can be found in numerous ways, including 'evidential value' (that certain places 'yield evidence about past human activity'),[4] 'historical value' (that some places link us to the past), 'aesthetic value' and 'communal value' (the meanings of places for people 'who draw part of their identity from it, or have emotional links to it').[5]

The connections between the three circles of sustainability can be clearly seen in the ways social/cultural benefits express themselves in monetary terms. The *Guide to the principles of the conservation of historic buildings* notes that a building can have economic value not just as a result of some sort of function, but also 'indirectly, in that its character, quality, interest or beauty enhances the value of the immediate area in which it is set, or of the wider area or country as a whole'.[6] Some notion of value can be estimated using hedonic pricing studies – an attempt to understand how value is signalled by the willingness of people to pay to retain heritage assets, but that often relates to monumental historical structures, such as cathedrals and large country houses. However, the value of some major asset, perhaps a cathedral, may provide a web of economic value which is not realised by the asset itself or by the people who visit it, but is distributed over the immediate environs and the nation – or perhaps the world as a whole. The value of historic streetscapes in some areas might be expressed in terms of enhanced rents. However, that may only be part of the overall value. The worth of un-monumental, 'ordinary' older buildings' architecture is not easy to quantify, nor is the collective worth of ordinary older neighbourhoods. Few architects would attempt such an assessment.

The complexity of the sustainability concept is easily seen in decision-making relative to older buildings and neighbourhoods – certainly in relation to the multiple dimensions that emerge from the Brundtland Report. *Conservation Principles, Policy and Guidance*, in an analysis of sustainability, cautions us about 'the difficult task of anticipating the heritage values of future generations, as well as understanding those of our own'.[7] Values are not absolute: before the 19th century there was little interest in the historical built environment. We can observe changing trends – some Victorian buildings once regarded as horrific are now held in considerable esteem. Again, it is difficult to predict future trends.

11.3 Looking at One Heritage Building Quandary

One might well consider Blackwell, in the Lake District, a house designed by the Arts and Crafts architect MH Baillie Scott and completed in 1900, as a building that is worthy of preservation. After a number of owners and use as a school, it had an uncertain future, but was purchased by the Lakeland Arts Trust in 1997. Restoration was undertaken using a grant of £2.25 million from the Heritage Lottery Fund, as well as proceeds from fundraising efforts. That Blackwell has significant social/cultural value was confirmed by the money contributed by social agencies and private individuals.

Part of the building was turned into an exhibition space, including the addition of mechanical air conditioning. The addition of climate-controlled exhibition space will consume more energy, but that inefficiency is offset by its social/cultural value.

In the case of such a building, window replacement and wholesale upgrading of insulation would undermine the very reasons for the preservation.

The presence of older buildings, encompassing listed monuments and unlisted, un-monumental buildings, together with the everyday older stock, necessitates a closer look at what comprises overall sustainability. Buildings are rarely unchanging, and improving energy efficiency is something that can be accomplished through appropriate refurbishment. Most importantly, however, regard has to be had for the social/cultural benefits that buildings can offer. Our urban areas are scarred by the unfortunate results of past attempts to undertake large-scale replacement, and the outcomes have often been socially disastrous. The problem of identifying, understanding and quantifying social/cultural value is one reason why it is difficult to reconcile older buildings with environmental and economic factors. As a result of social/cultural value, true sustainability may imply preserving some buildings or neighbourhoods in perpetuity, regardless of the environmental costs.

11.4 What Can We Learn from Older Buildings About Energy Efficiency?

While it is popularly believed that older buildings are heavy consumers of energy, there is evidence that this is not always the case, so such buildings must be dealt with on a case-by-case basis. A study undertaken by the UK Ministry of Justice/HM Courts Service looked at the energy use in 256 buildings. While their portfolio includes a variety of building types and patterns of use and management, the results were clear. The oldest buildings (pre-1900) had the lowest energy consumption, with 20th-century buildings built prior to 1960 having progressively higher energy consumption. After the 1960s things improved, until eventually: 'The court buildings of the 1990s and 2000s have managed to almost equal the energy efficiency of the pre-1900 buildings'[8] The study also found that among the buildings built in the 1990s and 2000s, the larger buildings were the least efficient.

Studies prepared by engineering firm Morrison Hershfield for the Athena Institute compared the performance of refurbished older commercial buildings with that of new buildings in Canada, where many cities encounter extreme climate conditions. For example, Winnipeg experiences hot, humid summers (the maximum daily temperature is frequently over 30°C) and long series of winter days with temperatures below -20°C (the average daily low temperature at the Winnipeg airport in January is -22°C), and has seen temperatures as low as -42°C. They found that refurbished office buildings could be more energy efficient than new ones, largely because of their smaller windows. In this case, an older construction tradition which could not rely on extensive mechanical systems created buildings responsive to a specific, and severe, climate – even for office buildings of eight to twelve storeys. They offered the comment: 'The most notable finding appears to be that the physical constraints of a heritage building do not appear to limit the potential for energy performance of a building. The biggest limitation may be the level of intervention for the renovations of the historic building.'[9]

This is echoed by Anne Power of the UK Sustainable Development Commission, who noted, with regard to the UK housing stock, that the oldest elements are 'often the easiest to renovate and make more efficient. Thus, there is almost an inverse relationship between the scale of current decay and neglect and the recycling potential of an area.'[10] Power also saw that new construction typically uses several times the amount of energy as refurbishment, and laments that calculations supporting widespread housing replacement often ignore the embodied energy involved in building replacement, the large amount of waste resulting from demolition and the social disruption incurred by forced rehousing.

Studies such as these suggest that thoughtful refurbishment of older buildings can lead to high levels of energy performance, and also that new buildings should embrace the performance-enhancing aspects of the older stock. The UK court system, as a result of its studies, developed a set of guidelines, including bringing back into use 'Redundant and underused space in pre-war and pre-1900 buildings (such as attics, basements and outbuildings)'. Their departmental architect noted: 'This is not only for their historic, aesthetic and cultural importance and the embedded energy, but also for continuing energy conservation.' New court buildings are to include operable windows, natural ventilation systems, natural lighting, and curtains – features common in older buildings.

The point is that wholesale condemnation of an older building stock, based on collective numbers, is inappropriate – buildings have to be approached on a case-by-case basis using data specific to the case, by designers who are sensitive to the social/cultural values associated with such structures. When looking to maximise societal returns from investments to advance a comprehensive sustainability, the older building stock offers potential that should not be ignored.

11: Notes

1   *Guide to the principles of the conservation of historic buildings*, BS 7913:1998, p. 7

2   David Pearce and Edward B. Barbier (2000), *Blueprint for a Sustainable Economy*, London: Earthscan, p. 20

3   English Heritage (2008), *Conservation Principles, Policies and Guidance for the Sustainable Management of the Historic Environment*, London: English Heritage, p. 27

4   Ibid., p. 28

5   Ibid., p. 31

6   Ibid., p. 6

7   *Conservation Principles, Policy and Guidance,* p. 46

8   John Wallsgrove (2008), 'The justice estate's energy use', *Context*, 102, http://www.ihbc.org.uk/context_archive/103/wallsgrove/page.html, accessed 12 August, 2011, p. 19

9   Athena Sustainable Materials Institute and Morrison Hershfield Limited (2009), *A Life Cycle Assessment Study of Embodied Effects for Existing Historic Buildings*, www.athenasme.org/publications/docs/Athena_LCA_for_Existing_Historic_Buildings.pdf, accessed 20 June, 2011, p. 22

10 Anne Power (2008), 'Does demolition or refurbishment of old and inefficient homes help to increase our environmental, social and economic viability?', *Energy Policy*, 36, p. 4, 489

When Things Become More Complex: Future Decisions and Flexibility

12

It is very tempting to collect a pile of data, force it into a model (a 'black box'), turn the handle and see an accept/reject answer appear. Unfortunately, even though the answer might be technically 'correct', the process may fail to incorporate some important characteristics of the alternatives being considered. One significant problem is that discounting methods were originally designed to analyse essentially passive investments, such as bonds, so may fail to reflect the fact that 'real' projects are subject to decisions made by people in the future, inevitably using information that is not available to you. In business this is quite a normal situation. It is unusual to make all the decisions for the entire life of a project at the beginning.

12.1 Decision Trees

Decision trees are useful when addressing situations which may involve sequential decisions, which are cases when one decision leads to an opportunity to make another. While the concept is based in the business disciplines, they can help to illuminate the environmental and social/cultural dimensions too. For the designer they have the added attraction that they create a visual picture of risks and rewards associated with different courses of action, and can be created and manipulated using vague information.

12.1.1 Mr Sullivan's development

Sequential decisions abound in the property world. Such situations often arise in conjunction with the possible development of vacant parcels of land.

Mr Sullivan is an experienced regional developer. He is considering building a retail project on the edge of a medium-sized city, and new growth is expected in the area over the coming years. The site will allow the construction of 8,000 square metres of retail space. How the demand will materialise is uncertain: there is local concern about urban sprawl, so it is not clear how much new housing will be allowed in the area, or how soon it might happen. Mr Sullivan likes to put projects together, build them and then sell them on to an investor group at a price based on the rent achieved. He has discussed the market with estate agents and surveyors, has talked to potential tenants and has managed to get conditional lease commitments for 1,800 square metres, but remains unsure of how things might ultimately unfold. If demand is low, the project will experience low rents and vacant shops. If demand is high he will be very happy, and especially happy if he has chosen to build the entire project. He wants to avoid building too big a project, and perhaps taking a number of years to lease it out, thereby giving the project a sense of 'un-success' which may taint it in the marketplace for years. He is pleased that he is not going to play 'bet your company' as he often did when he was younger; now, even in the worst-case scenario, he will not fail financially.

In discussion with the local planners and his bankers, Mr Sullivan has settled on two possible alternatives. The entire project can be built at once, or in two phases of 4,000 square metres each. He does not want to keep his capital tied up for too long in this project, so he will proceed very soon with the second phase. If the rental income is not going to materialise within a couple of years, he will sell the project for what he can get, even if it means taking a loss, and move on to his next venture.

Mr Sullivan starts to put together the material to enable his decision. The fundamental question is whether to build the project all at once, or in two phases. His architect has done a sketch design of each alternative, and some rough cost estimates have been prepared. The two-phase project will cost more. Other associated questions arise: does the cost of building two phases outweigh any advantages of that approach? Are other opportunities being lost by building the project in phases?

Decisions such as these are difficult to evaluate using the usual discounting methods because of the second-stage decision.

Creating a decision tree will help Mr Sullivan appraise the situation. He recognises that decision trees are somewhat crude representations of reality, in common with most decision-making aids. Yet it will force him to collect information and to give consideration to the most important factors in determining whether or not the project will be remunerative.

After completion, his intention is to sell the project to an investor group. It will be possible to assess the project's value by means of discounting, and that is essentially what Mr Sullivan's purchaser group will be doing when they apply a 'cap rate' (capitalisation rate) to the attained net rental income (after all expenses) to arrive at a purchase price. As was seen in Chapter 6, this is a simple way of assessing a perpetual cash flow, or in this case, one that should continue for quite a number of decades.

Mr Sullivan has decided to explore three possible levels of demand, something that will be determined by some development approvals in the geographical area. He has assembled his data in three tables.

Figure 12.1 Demand and possible sale price of project					
	Net rent: €/Sq m	Sq m	Net rent	Cap rate (%)	Sale price (€)
High demand		10,000 (Limited by buildable area)			
Full size	400	8,000	3,200,000	6.5	49,230,769
Phase I only	400	4,000	1,600,000	6.5	24,615,385
Medium demand					
Full size	320	8,000	2,560,000	6.5	39,384,615
Phase I only	320	4,000	1,280,000	6.5	19,692,308
Low demand					
Full size	221	3,600	794,880	6.5	12,228,923
Phase I only	221	3,600	794,880	6.5	12,228,923

Development cost		€/Sq m	Sq m	Total cost (€)
One phase		3,600	8,000	28,800,000
Two phases	Phase I	3,942	4,000	15,768,000
	Phase II	3,745	4,000	14,979,600
			Total two-phased project	30,747,600

Outcomes		Sale price (€)	Cost (€)	Net profit (€)
Build complete project				
High demand		49,230,769	28,800,000	20,430,769
Medium demand		39,384,615	28,800,000	10,584,615
Low demand		12,228,923	28,800,000	-16,571,077
Phased process				
High demand	Build phase II	49,230,769	30,747,600	18,483,169
	Do not expand	24,615,385	15,768,000	8,483,169
Medium demand	Build phase II	39,384,615	30,747,600	8,637,015
	Do not expand	19,692,308	15,768,000	3,924,308
Low demand	Build phase II	12,228,923	30,747,600	-18,518,677
	Do not expand	12,228,923	15,768,000	-3,539,077

Simply laying out the alternatives is revealing: it confirms that phasing the project eliminates the possibility of a major loss. Mr Sullivan realises that if there is a medium demand, he might be able to lower the rent somewhat to achieve full occupancy. It is obvious that to achieve the greatest profits, the larger project has to be built, but he also recognises the uncertainty of demand: while things look good now, it will take two or three years before the project is complete, and the economy (both global and local) could change significantly during that time. Some of the alternatives are clearly dominated by others. For example, if only one phase is built and high demand is experienced, it is only logical to build the second phase. If there is medium demand it does not make sense to build the second phase – at least not immediately. A simple sketch of the decision helps him understand the various possible paths.

12.1.2 Creating the decision tree

The decision-making process can be started by drawing a decision tree. One starts on the left-hand side of the page; today's decision has to be made in the present. For each alternative a branch is drawn to the right. From there, future outcomes and decisions are laid out to create more branches. These are fairly obvious in the case of simple trees such as that being drawn by Mr Sullivan. The usual convention is that decision points are represented by squares or diamonds, while points of uncertainty are represented by circles. The first point is called the 'root node'. Earlier, opportunity costs were discussed: every decision point, especially at the beginning of a project, actually contains a myriad of alternative possibilities. There is always the possibility of building a hotel or office building on the site, or not building anything and putting the money in the bank. This is why any feasible decision tree is only a sample of possible paths.

The decision-maker should reflect on the bare tree (Figure 12.2) before appending data. A number of questions might arise at this stage. What are the elements of uncertainty that might affect the outcomes? Have all the significant decisions and possible outcomes been properly reflected in the tree?

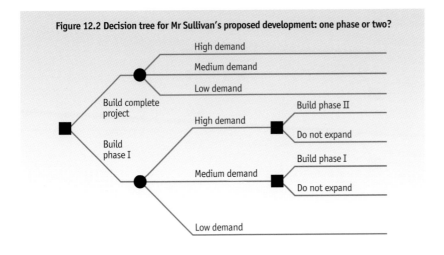

Figure 12.2 Decision tree for Mr Sullivan's proposed development: one phase or two?

High demand

Medium demand

Low demand

Build complete project

High demand

Build phase II

Do not expand

Build phase I

Build phase I

Medium demand

Do not expand

Low demand

12.1.3 Decision trees and numbers

At this stage data has to be assembled and put on the tree. Mr Sullivan has already collected the key pieces of information (Figure 12.1). He has identified one interesting issue: his research suggests there may be a demand for as much as 10,000 square metres of space; however, he is limited by the site area and the nature of his planning permission. He cannot secure all of it, but if high demand materialises and he only builds one phase, some other developer might decide to build another project. Of course, having completed a first phase will be an advantage in attracting tenants for a second phase. Mr Sullivan thinks this through and believes that if the high or medium levels of demand materialise, he can move quite quickly into his second phase; after all, he has planning permission and will have already built much of the site infrastructure. That ability to expand quickly has value. If Mr Sullivan only built the first phase and then sold the project, he would receive an amount based on the return from the first phase *plus* something reflecting the opportunity to build more. There is lots of room for speculation, but Mr Sullivan can start exploring the decision with the data he now has.

A few numbers might be put on the tree, including costs, sales prices and profits at the end of each branch (Figure 12.3). For Mr Sullivan these are all the costs associated with developing the project, including the land (because a do-nothing branch, regardless of whether it is drawn or not, ultimately results in the sale of the land). There is some time involved, so the time value of money should be accounted for, however, given the complexity of real-estate finance, and in the interests of keeping the analysis reasonably transparent, this might be ignored in a first evaluation. At each point of uncertainty, the user assigns a probability for each outcome (they total 100 per cent). As might be expected, these tend to be managerial best guesses, and are simple projections expressed as two or three possibilities.

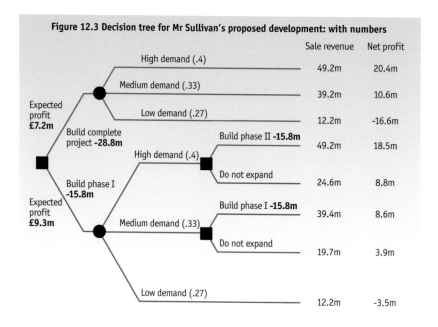

Figure 12.3 Decision tree for Mr Sullivan's proposed development: with numbers

	Sale revenue	Net profit
High demand (.4)	49.2m	20.4m
Medium demand (.33)	39.2m	10.6m
Low demand (.27)	12.2m	-16.6m
Build phase II -15.8m	49.2m	18.5m
Do not expand	24.6m	8.8m
Build phase I -15.8m	39.4m	8.6m
Do not expand	19.7m	3.9m
Low demand (.27)	12.2m	-3.5m

Expected profit £7.2m

Build complete project -28.8m

Build phase I -15.8m

Expected profit £9.3m

High demand (.4)

Medium demand (.33)

12.1.4 A decision tree solution

A solution can be obtained by thinking about what decisions would logically be made in the future. In this case one starts with the right-hand side of the tree (the final possible states), and works back to the present.

At each uncertainty node the result is the sum of the probability-weighted outcomes (the net profit) of the different branches (multiplication of the value of the outcome by the associated probability). This number is the expected value of the outcomes at that node.

Starting with the 'Build complete project' alternative, an expected outcome can be computed, covering the three branches that reflect differing levels of demand. The outcome at each demand level is multiplied by the probability of experiencing that demand, and the three added together.

Expected outcome = (probability of high demand × outcome under high demand) + (probability of medium demand × outcome under medium demand) + (probability of low demand × outcome under low demand)

so, in this case:

Expected sales revenue	= (.4 × 49.2) + (.33 × 39.4) + (.27 × -16.6)
	= £36.0 million
Less development cost	(28.8 million)
	= £7.2 million

A similar process is undertaken for the phased alternative. At each decision node the cost of each alternative branch, less the cost of following that branch, should be noted – and the branch with the highest benefit should be selected. In this case, some can be selected on the basis of inspection because the best route is obvious. If there is high demand one should complete the second phase, and if the demand is medium or low it is best to build the first phase only.

In this fashion, the decision-maker moves from right to left, ultimately arriving at the point of the original decision, at which the values of the initial alternatives can be compared. It can be seen that the analysis suggests building the project in two phases.

Looking at the phased development, it is obvious that if high or medium demand is experienced the second phase should be built; its net profit is higher than the 'do not expand' outcome, so it would naturally be chosen. But if low demand is experienced expansion would not be undertaken.

Bailing Out

The ability to abandon a project can add significant value to it as an alternative. Property development is a relatively easy process to abandon, as long as it is done early enough, because usually land value can be realised through sale. As the process proceeds though, certain costs become difficult to recover. Land value is usually enhanced by the acquisition of planning permission, but additional architectural work and marketing costs associated with acquiring tenants may have to be written off. One interesting abandonment possibility is bankruptcy. Investors in corporations are usually personally protected, unless they make explicit guarantees. While they might lose all the money they have invested in the project, they have the possibility of turning the mess over to the courts and the creditors. This structure allows many projects to proceed that otherwise would not have been possible. Investors are more likely to contribute equity money when they know that they are not risking their entire worth. This is one of the reasons why so many real-estate projects are undertaken by corporations with few other assets. Not surprisingly, few proposals to lenders outline the possibility of simply walking away from a project.

12.1.5 Considering the meaning/dealing with the output

The output of this process suggests that Mr Sullivan would be advised to build the first phase and see what happens. The expected profit for the phased approach is £9.3 million, in contrast to the £7.2 million if he were to build the entire project at once. If either medium or low demand occurs, he will be better off. The possibility of a big loss has been eliminated, although if low demand occurs the phased project will incur a loss of £3.5 million. It is likely that Mr Sullivan will sense that this analysis is incomplete, and might elect to take it to a further stage. Having a high demand in the first period does not ensure ongoing high demand, although it is probably an excellent indicator. Moreover, if only one phase is built there will be some value remaining in the unused land (perhaps more than £3.5 million), and other possibilities abound. He might choose to build a hotel or some offices.

As with all decision support mechanisms, decision trees are only rough representations of reality and contain a number of limitations. That will be obvious to anyone who has plotted one out.

- Decision trees can become complex very quickly. With regard to Mr Sullivan's decision, there will always be more factors and possibilities demanding to be included in the analysis. Decision trees are best used to understand linkages between key factors, and to understand how decision-makers, both now and in the future, can respond to them, so keeping them simple is best. A more complete analysis could be created with a Monte Carlo simulation. Although specialised software is available to create and analyse decision trees, there is much to be said for the pencil and paper approach, at least for initial explorations.

- Most real business possibilities are continuous distributions, ranging from the wonderful, through the most likely, to the truly disastrous. Decision tree analysis requires that the user simplify these into a small set of discrete possibilities, typically 'good' and 'bad', and possibly one or two intermediate possibilities.

- Decision trees usually rely on subjective probabilities created by the decision-maker, and these may be only best guesses, which depend upon the capabilities and biases of the analyst. Better calibration, in accordance with any available data, is desirable.

- Decision trees can only incorporate the paths envisaged at the time of the initial decision. Future decision-makers may have alternatives that cannot be discerned now, or that are not yet available. Emerging technology and markets can create new paths.

- The construction of decision trees necessitates the inclusion of some less-than-stellar outcomes, and those are often difficult to present to senior managers, lenders or others whose support may be required to further a project. One can only imagine the reaction of a lender or equity investor to hearing the details of a bail-out strategy. An individual decision-maker might put the bail-out branch on a private decision tree. Many managers don't want to hear about the downside scenario.

- Sensitivity analysis might be undertaken to understand the nature of the different alternatives. Mr Sullivan, for example, might be interested in knowing how changing the demand probabilities could affect the answer.

12.2 Sustainability: Two More Trees?

As discussed, the first step of a decision tree analysis is to create a tree of possibilities, mapping out present and future decisions and possible outcomes. Only as a second stage are values and probabilities attached to the various branches. The reason that the data is initially economic is because of the assumption that future managers will continue to make decisions primarily on an economic basis, even though there are accompanying environmental and social/cultural implications. Regardless of this, decision trees can help the designer or manager visualise the relationships between the different aspects of sustainability. Decision trees are usually laid out in two dimensions, constrained by a piece of paper or a computer screen. However, one might consider them in three-dimensional space, one atop the other (Figure 12.4). The paths and branches would be the same, and the same probabilities would be attached to each branch, but the outcomes on each level would be measured in different terms.

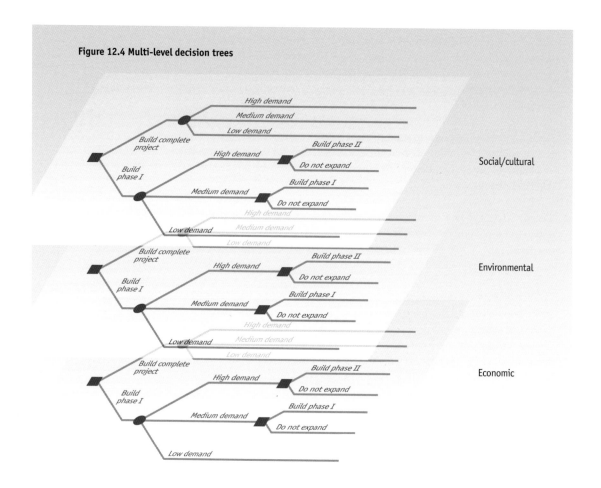

Figure 12.4 Multi-level decision trees

The complexity of this should be immediately apparent, and also that specific predictions of the environmental consequences of design and development decisions are almost sure to be wrong. Events may unfold in the future which will be responded to by managers, and the course of the project will change as a result. Will future decision-makers put more weight on the environmental and social/cultural dimensions? Possibly, although these do also spill over into economic aspects of projects.

There are many possible implications associated with each branch and outcome, and one would immediately consider such issues as energy consumption, CO_2 emissions and water and air pollution. There are also social and cultural implications. This multi-layered decision tree way of thinking may provide the decision-maker with a better view of the alternatives. One of the good features of this analysis of environmental and social ramifications is that it requires that the decision-maker consider alternative paths.

While mapping out future possibilities on decision trees has some benefits, a familiar problem emerges: while most business decisions are made on the basis of a single indicator (money) and a single tree will suffice, environmental and cultural/social matters are complex in themselves, so conceptually might warrant a virtually infinite number of layers. Nevertheless, a tree analysis of energy consumption or CO_2 emissions will help the decision-maker avoid overinvestment or underinvestment because of the recognition that there are ranges of possible futures.

Also, there is a further level of complexity. In some cases it will be reasonable to attach the business uncertainties to the environmental and social/cultural dimensions. Yet the presence of other uncertainties means that this should not be done without some reflection. Non-economic decision trees need to be used with caution as a decision-support system, and not as a complete decision-making system in themselves.

The characteristics of Mr Sullivan's predicament cascade into project details. Starting assumptions were made about how much would be spent on a first phase, but would it be better to spend more (or less) on the site services? Would the design be fundamentally different for a large shopping facility than for a small one? What would happen to the design if other uses were attached to the 'low demand' branches? How might this change the materials and mechanical systems chosen?

12.3 Decision Trees as Tools

Decision trees are useful tools in cases when future decisions may be made. Mapping out the branches forces the decision-maker to analyse the situation thoroughly. With respect to uncertainty, it requires that the decision-maker state a quantified uncertainty, much as Mr Sullivan had to put figures on the probability of various levels of demand.

Peeling back the layers of the decision through more advanced analysis challenges the decision-maker to explore and understand the fundamentals of the situation, in particular with respect to uncertainties, and then apply some mathematics to ensure that the decision is in keeping with those essential facts or informed beliefs.

Again, sensitivity analysis is a useful tool when added to decision tree investigation. In particular, changing the assumed uncertainties will help one to see how the suggested course of action might change. Mr Sullivan might well ask, 'If the probability of high demand was higher, would the decision change?' (And how much higher?)

The identification of the sources and the nature of uncertainties can help to identify opportunities and appropriate responses, now and in the future, helping decision-makers to exploit upside potential while reducing downside exposure. Empowering future decision-makers to be able to respond appropriately to emerging conditions is a way of adding to overall project value.

Building Lives into the Future: Uncertainty, Adaptability, Flexibility and the Real Options Concept

13

13.1 Incompleteness of Tools

In an earlier chapter it was apparent that Catharine Napier's analysis of flooring alternatives was somehow incomplete. Data is missing – sometimes because it does not, and perhaps cannot, exist, and available analytical tools seem to leave their users feeling uneasy. In research conducted by the authors, one reason for the under-utilisation of whole-life costing techniques expressed by experienced designers and decision-makers was that the results frequently did not accord with their instincts.

Often, the changing nature of the world is not fully reflected. In the case of Ms Napier's decision, the relative prices of the different materials can change, the materials themselves can change, wear is uncertain, design fashions can change, and the future strategic directions of the department store chain and the markets in which it operates cannot be known with much confidence.

Good insights can come from attempts to understand the nature of what is uncertain. Mr Sullivan's use of decision trees to decide whether to build his shopping centre in one phase or two improves the understanding of sets of alternatives.

Two recurring issues in conventional whole-life costing methods are incomplete recognition:
- of the implications of the uncertainty in which projects exist
- of how future decision-makers might react to various future situations.

There is a need for some degree of humility when making decisions about buildings. It is tempting to believe that the future can be forecast, and forecasts abound: about future energy prices, the future prices of investment assets (such as property), what might be in demand in the future, and the nature of the environment. The reality is that many forecasts do not come true. If one regards outcomes as probability distributions, it is highly unlikely that any specific projection will be precisely on target.

There are reasons why publicly available forecasts are likely to be wrong. Anyone with a good record of forecasting would be wise to keep information secret and exploit it in the stock market or at the racetrack, because letting others know might compromise one's advantage. A second reason is that forecasts based on the projection of recent trends may be ignoring the fact that many ongoing processes have some tendency to return to a mean. The reasons for this 'mean reversion' are logical. If, for example, some commodity experiences an extreme increase in price, new sources of supply may appear and consumers may limit their consumption and attempt to substitute some other good or service, thereby pushing the price back down. If there is a shortage of hotel rooms, for instance, someone will probably build some.

To further upset any confidence one might have in long-term projections, established trends are often broken by unpredictable events, sometimes described as 'wild cards' or 'black swans'.[1] Some are easy to identify – certainly the 2001 attack on the World Trade Center is one – and periodic stock market crashes are another, as are political crises, new technologies and major upheavals in lifestyle. Such events are important, because most analysis, including much done in this book, has its roots in statistics, and those 'once-in-the-life-of-the-universe' events (that may have happened three times in the last month) fall outside normal analysis.

This leads to methods that augment the discounting-based methods' approach of setting an 'expected' set of future events and using it as the base for the analysis. The reality is that the path travelled by any project may be very different from that upon which the original analysis was based.

Over the past thirty years a set of 'add-on' approaches have been developed, which can move the whole-life costing exercise closer to revealing the 'right' answers.

13.2 Flexibility is Good. How Much Should You Pay for It?

13.2.1 Enhancing value: flexibility and real options

In 1994 Stewart Brand created an illustrated book called *How Buildings Learn: What Happens After They're Built*; it became popular, and a television series followed. It is difficult to refute his essential claim – that flexible, easy-to-modify buildings are a good idea. This is also one of the themes presented by Jane Jacobs' *The Death and Life of Great American Cities* (1961): that prosperous, sustainable cities have to offer the easy-to-modify sorts of spaces Brand suggested.

So, everyone knows that flexible buildings are a good idea, but neither Brand nor Jacobs tells us how much to pay to get flexibility. Of course, in some cases, flexibility costs nothing, but in other cases there is some cost implication, either in terms of initial cost or of some compromise in the efficiency of the original use for which a building is being created.

It is clear that the composition of value is not a simple concept, and that it varies from time to time and from place to place. How future inhabitants of our societies might assign value cannot be known. Social and economic change may mean that some buildings lose all of their value and ultimately are demolished. This is one aspect of the uncertainty that dominates decisions which are to be measured against future benefits, and it is the reason why sustainability demands that attributes be included in buildings to increase the probability of their survival. However, to make detailed plans for future generations would be as ridiculous as the Victorian builders making provision for our needs. Although we do use much that the Victorians left us, we are effectively picking through their leavings to find what we can use, rather as a vagrant searches through rubbish tips.

As reflective decision-makers, it is reasonable to aspire to decisions that give future decision-makers favourable situations and resources with which they can make their own decisions. In a building context, this can be done through flexibility. Victorian factories and warehouses can serve us now because of their robust construction and reasonably open spaces. It is fair to wonder whether our appreciation for their design could have been predicted, but as we have seen, for many decades Victorian buildings were prime targets for demolition.

Sometimes boarded-up buildings are kept for significant periods of time and are subject to only minimal amounts of maintenance by their owners. The owners must perceive some possibility of future use, otherwise they would not keep them. Properties are abandoned only when, relative to the carrying cost, any hope of reasonable use in the foreseeable future has disappeared. The more flexible they are, the more likely they are to be used for something in the future, so the more likely they are to survive. An owner has an option to do something: the right, but not the obligation, to undertake some use, perhaps a redevelopment, of the property. Mr Sullivan, our shopping centre developer from Chapter 12, by constructing only part of his project, retains the right to build the rest later, if conditions so warrant. It depends upon how rents move in the future.

The managerial disciplines term this form of value as resulting from the presence of real options. Real options analysis is now an accepted way of making corporate decisions. It helps to resolve some of the basic issues in usual discounting practices, in particular the use of expected resource flows, and the implicit assumption that projects evolve predictably.

Quantification of real options emerged from the field of traded financial options (typically 'puts' and 'calls'), which was revolutionised through Nobel Prize-winning work by Myron Scholes, Fischer Black and Robert Merton in the 1970s. Although various forms of options had been traded for thousands of years, the ability to compute their value was limited. Since the 1970s, a variety of methods have been developed, and the study of options (both financial and real) and offshoots has become a major activity of business schools.

Copeland and Antikarov offered a definition: 'A real option is the right, but not the obligation, to take an action (e.g., deferring, expanding, contracting, or abandoning) at a predetermined cost called the exercise price, for a predetermined period of time – the life of the option.'[2] This implies that there are one or more decisions to be made in the future; that the course of the project is not 'set in concrete', and that it can respond to future events. They suggested 'that NPV (Net Present Value) systematically undervalues every project', but a bigger problem is that net present value calculations alone may undervalue some projects more than others, thereby possibly leading to an incorrect ranking of project alternatives. Generally, flexible projects that allow future decision-makers to adapt to new requirements will have a higher value than those that do not.

Given the origins in the financial community, it is worth explaining that the holder of a call option, one of the basic forms, has the right to buy an asset, perhaps the share in a corporation at a specific price, during a specified time window. If at the 'exercise date' the value of the share is over the 'exercise price', the option owner wins; if the price is lower the option is not exercised and disappears. As the option has to be purchased, if it is not exercised the owner has lost the purchase price.

Options can be risky – they may not be used, but they can lead to significant returns if certain conditions unfold and they can be used. Many corporations use them to manage risk. Drummond states that options 'are about taking affordable risks that help avoid damaging losses and may lead to substantial gains', and calls them 'toehold investments'.[3]

So, how can the construction or property professional apply this in practice? Options abound in the property world. Some come with labels, such as 'option to purchase', but most don't. Making provision to expand a building or change its use are options. Oversizing building services is an option where one pays more now in order to be able to increase the size of the building without digging up the site. There will be instances when expansion never occurs, but that does not mean that paying the extra amount was wrong. An affordable risk was taken that might lead to those substantial gains – the case when expansion occurred without requiring digging up the site to replace or upgrade the services.

One option is the ownership of a vacant piece of land, which can be compared with a financial call option (Figure 13.1).

Figure 13.1 Comparison of financial and real options associated with vacant land		
	Financial option	**Vacant piece of land**
Underlying asset:	Underlying asset (stock)	Some type of building
Cost of using option:	Exercise price	Cost of development
Profit:	Share price minus exercise price	Value of building at completion less development cost
Time frame:	Specified	Usually unspecified
Cost of acquiring option:	Option price	Land value

The same characteristics that make property-based decisions difficult in general apply to real options decision-making. Moreover, the mathematics associated with real options can become complex quite quickly, so most practitioners will avoid many of the associated techniques. But the good news is that a basic understanding of how real options work and being able to identify when they may be important will yield dividends. In particular, a real-options way of thinking means that uncertainty is accepted as something to be understood, managed and exploited.

13.2.2 Looking long-term: things can change

Conventional whole-life costing tends to assume ongoing like-with-like replacement. However, everyone knows that choosing cheap carpet this year does not mean that it will be selected again next time. Decisions will be made in the future that affect the course of the buildings that we are creating (or managing) now – often in unpredictable ways. If we give future generations more resources and more latitude for their decisions, our buildings will be more sustainable. This extra worth can often be quantified, but even just laying out and thinking about the different possibilities will help to inform the decision-maker.

Figure 13.2 shows the view of the future implied by a typical whole-life costing: the initial decision is repeated into the distant future with, in this case, cheap carpet being replaced by more cheap carpet, and rubber tile with more rubber tile. The decision-maker has assumed that today's choice is good for all of time.

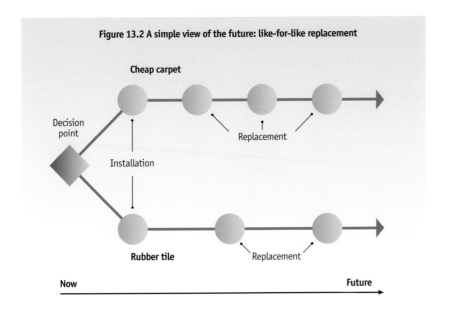

Figure 13.2 A simple view of the future: like-for-like replacement

Cheap carpet

Decision point

Installation

Replacement

Rubber tile

Replacement

Now

Future

Reality, though, is more complex (Figure 13.3). Different choices may be made in the future. While this sketch shows only two alternatives, in reality a large set of possibilities may be available at each step.

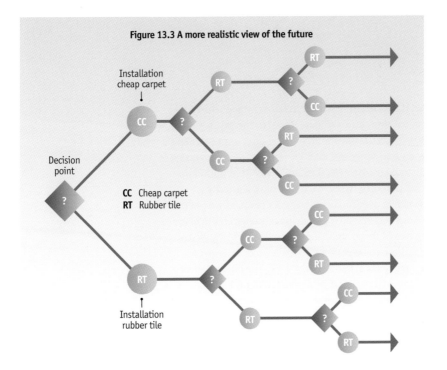

Figure 13.3 A more realistic view of the future

Installation cheap carpet

Decision point

CC Cheap carpet
RT Rubber tile

Installation rubber tile

An even more realistic illustration would take into account the likelihood of new products becoming available for consideration and possible selection when the decision is made again (Figure 13.4). When the cheap carpet wears out, a decision-maker may decide to use some new material that does not now exist – or a product that exists now but has become relatively cheaper.

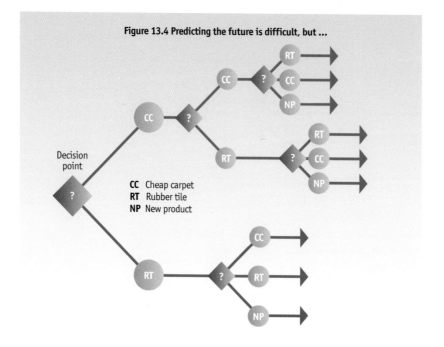

Figure 13.4 Predicting the future is difficult, but ...

If we look at Ms Napier's carpet decision again, we might see how and why the value may differ between the initial alternatives (Figure 13.5). In the assessment of value, short-life alternatives, such as the cheap carpet, may carry some element of value that the long-life ones do not. While the long-life and low-maintenance requirements of terrazzo might appear attractive, they are dependent upon long-term survival to achieve their full potential. While we might have a good sense of what might happen to the building over the next five years or so, as we peer into the more distant future things become increasingly uncertain. In particular, the likelihood of encountering one of those annoying wild cards or black swans increases. It is not necessary for these to be massive, global events. If after a few years the management decide they need a dramatic new look, the cheap carpet can be torn up and replaced with few regrets, while replacement of the long-life materials would mean that their full value was not realised – and therefore that the long-life material was the wrong choice.

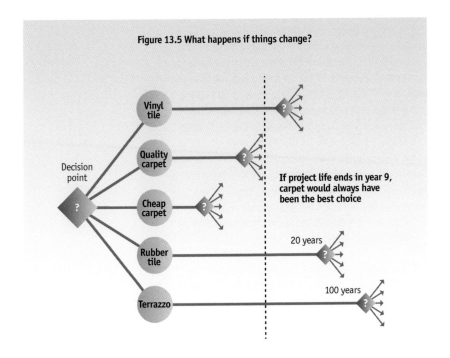

Figure 13.5 What happens if things change?

Decision point

Vinyl tile

Quality carpet

Cheap carpet

Rubber tile

Terrazzo

If project life ends in year 9, carpet would always have been the best choice

20 years

100 years

Traditionally, the message of whole-life costing has been to take a long-term viewpoint – contrary to what so many decision-makers seem to do intuitively. Yet, when augmented with additional insights, whole-life costing sometimes suggests that cheap, short-life solutions may be appropriate. In particular, fashion or technological change may shorten the time period over which expected returns are received – for example, wireless systems have diminished the value of raised access floors. One of the tasks for the designer or manager is to determine when to take a long-term perspective, and when to make decisions for the short-term.

13.3 More Issues: the Energy Paradox

A good example of the influence of real options has to do with energy. While energy is one of the measurable aspects of 'green buildings', there has often been a reluctance by consumers and businesses to make energy-saving expenditures unless they expect to receive very high returns.[4] Stastad et al. (1995) reviewed various papers which revealed implicit discount rates for energy efficiency investments ranging between 25 per cent and 300 per cent, with typical rates being 30 to 40 per cent, corresponding to paybacks of between 2.5 and 3.5 years. This expectation forces governments to offer subsidies to encourage reluctant consumers and businesses, or to legislate for energy-saving requirements such as higher insulation standards for new construction.

The problem is that while actual in-service energy use can be quantified, before a building is built it is more difficult to understand what the value of energy-saving features will be. This issue leads to what has been termed the 'energy paradox', and has been explored at least as far back as the early 1970s. Dr David Fisk, then of the Building Research Establishment, in a 1976 paper saw such behaviour as making sense.[5] He thought that what he termed the 'economic optimum' could only be understood in the context of uncertainties with respect to energy prices. Why?

The return from investing to achieve energy savings, such as through additional insulation, better windows or more efficient mechanical equipment, is based on energy costs. The savings realised are greater if energy prices are high. However, if energy prices turn out to be low, you might regret buying them. It is also an irreversible decision: once an energy-saving investment has been made, one is at the mercy of the marketplace – the money is gone.

In this case, before the work is done, a call option structure exists, because retrofitting can be done now or in the future – or never. The building owner has the right, but not the obligation, to make the investment.

It should not be surprising that such option structures depend upon the volatility of energy prices – how much prices move around. Energy is a prime field for applications of real options techniques because there is precise data going back decades. Past movements of real gas and electricity prices (with inflation removed and prices indexed to 1990) are shown in Figure 13.6.

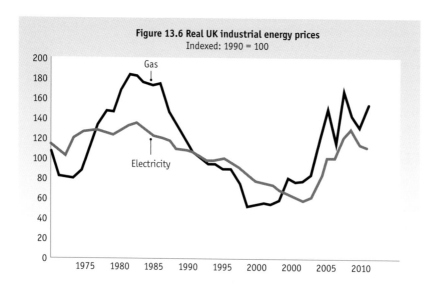

Figure 13.6 Real UK industrial energy prices
Indexed: 1990 = 100

Optimal decisions require consideration of this volatility. Anyone filling a car tank through the last few years will agree that the price of energy is subject to change. When petrol prices were at a recent high, newspaper articles argued that it made economic sense to sell a large car and buy something more fuel-efficient; the cost of making the change would be covered by the fuel savings. A few months later, if petrol prices fell, anyone who had taken that advice might be regretting the decision, having effectively overinvested. The return did not materialise and the expenditure was not warranted. The owner of the large car has the option, again the right but not the obligation, to buy a smaller car, and it can be exercised at any time until the large car disintegrates, at which point the option disappears, and the character of the decision changes. When that happens, the benefits of buying a fuel-efficient car can be calculated using basic discounting tools. So it is with energy-saving investments in buildings. If the decision can be postponed, it might be best to do so. After all, in the future we will have better information, so might be able to avoid making the wrong decision.

13.4 Modelling an Uncertain Process

What is interesting is that people apparently have an innate understanding of this type of situation while being unaware of the theory or the mathematical tools. How this happens is the subject of ongoing research, but this illumination of the process and explanation of the formal assessment methods may help the designer or manager make better decisions when these kinds of situations are encountered.

There are various means of assessing real options problems, some of which rely on complex mathematical techniques that are not likely to be immediately available to most designers or managers, and/or demand volatility data that may be unavailable. Less mathematical ways (not involving calculus) of assessing real options problems are more convenient and better suited to the world of buildings. One approach is to use binomial models, where uncertainty is modelled by coin flipping. Over a time period, perhaps three months, prices might go either up or down – heads or tails. Over a number of periods, a distribution termed a 'binomial tree' results from the series of coin flips. Past data indicates how much movement per period might be built into the series; Figure 13.7 shows how this works. Of course, the middle ranges are much more likely. They can be reached by a number of different sequences of coin flips, while the extreme ranges are dependent upon an unlikely series of only heads or only tails. After numerous periods the outcomes approach a continuous distribution.

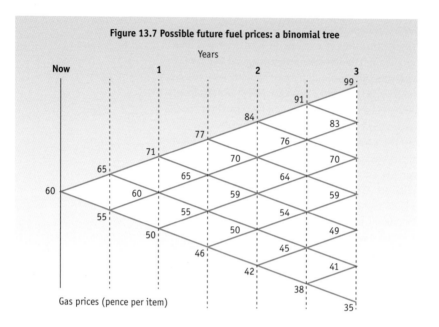

Figure 13.7 Possible future fuel prices: a binomial tree

Years

Now   1   2   3

Gas prices (pence per item)

Figure 13.7 helps to visualise the benefit of waiting. If the investment is not made, and energy prices go up, the investment can still be made. This is why waiting is often a good strategy, and this can be visualised quite simply, as in the second part of Figure 13.8. A very high starting energy price makes it less likely that prices will fall into a range of 'regret'. The high rate of return demanded by people making energy-saving investments is actually warranted. They wait until their investment will be reasonably protected from the implications of a fall in energy prices. Collective societal opinion may be different, perhaps being driven by environmental or energy security issues; hence the existence of subsidies or regulation to ensure that energy-saving investments occur.

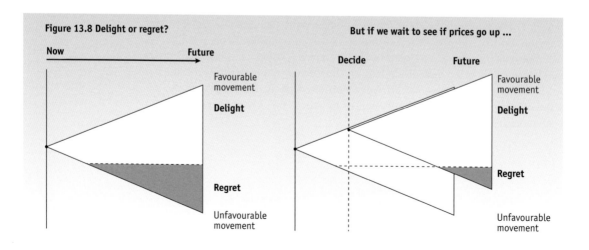

Figure 13.8 Delight or regret?   **But if we wait to see if prices go up ...**

Now   Future   Decide   Future

Favourable movement

Delight

Regret

Unfavourable movement

Energy works well as an example, because of the availability of good data, but there are many situations like this, and it is not always possible to collect the necessary data. A good rule of thumb is that when uncertainty is high and you can hold an option, waiting can be a good idea.

13: Notes

1   Nassim Nicholas Taleb (2001), *Fooled by Randomness*, London: Texere
2   Tom Copeland and Vladimir Antikarov (2001), *Real Options: A practitioner's guide*, New York: Texere, p. 5
3   Helga Drummond (2012), *Guide to Decision-making: Getting it more right than wrong*, Hoboken: Wiley, p. 162
4   Kevin A. Hassett and Gilbert E. Metcalf (1993), 'Energy conservation investment: Do consumers discount the future correctly?', *Energy Policy*, 21(6), pp. 710–16 and Alan H. Sanstad et al. (1995), 'How high are option values in energy-efficiency investments?' *Energy Policy*, 23(9), pp. 739–43.
5   David J. Fisk (1976), *Energy Conservation: Energy Costs and Option Value*, Watford: Building Research Establishment

Applying Real Options Concepts and Tools

14.1 Solar Orientation: Creating a Sustainability Option

Rick McInnes is the principal of a medium-sized architectural firm. One of the projects in his office is an urban, four-floor, forty-unit housing project. While Rick is very interested in solar applications, his client has reservations. The client tends to be focused on the 'bottom line', so Rick knows that he has to come up with convincing arguments to support his proposals. It is clear that the various alternatives now available in terms of photovoltaics or other forms of solar energy collection are not viable within the terms of reference of the client. But the client group is willing to listen, in part because they have a long-term perspective, and are always looking for ways to add value to their portfolio.

Rick sees the key as being the configuration of the roof and the adaptability of the heating system. A flat roof would be cheaper, but he would like to see the roof sloped and oriented for future solar collection. Accomplishing this has some cost implications, but they are not great. Rick wonders how he might assess this situation. He knows that he will get a favourable reception – the project manager recently read about artificial photosynthesis,[1] and is already wondering about very long-term possibilities.

14.1.1 Approaching an analysis

Rick is proposing to create an option that will allow for the adoption of any future solar-collecting technologies at low cost. He can work out both the extra costs and the savings in any future installation, but he does not know when, or if, the benefits of the roof alternative will ever be achieved.

This is like most flexibility options – they may never be used and, if they are used, one does not know when that might be. A classic example of this is movable office partitioning. Sometimes it is never relocated, and the original cost might be considered an outright loss. However, in other cases partitioning will be moved, sometimes a number of times, and the flexibility offered may be of considerable value. Flexibility can add value, so it may be worth acquiring even if it is never used. In that way it is rather like insurance – we rarely complain if we don't collect on insurance.

Solutions designed for financial options are difficult to use in this situation. In particular, there is no way of quantifying the level of uncertainty, because there is no history to consider. Rick should see that the benefit to be received from installing some unknown future technology will trigger the utilisation of his proposed option. The future return from having bought the option is the savings in not having to rebuild the roof. He might realise that one could model an option on emerging technology, perhaps by projecting the past improvements in photovoltaics into the future, but he recognises that as a mere architect he should be uneasy about making such predictions. Moreover, this option exists in a state of high uncertainty with respect to technology. While an innovation in solar energy collection might occur, causing the option to be exercised, it is also possible that some other breakthrough, – perhaps fusion-generated electricity – might cause it never to be used. Of course, things are never simple. If solar technology were easier and cheaper to install, it would be more likely that some future manager might decide to do this.

This sort of analysis can be used in many situations, with one being the design of site services. Is it best to size them for just the one phase, or to spend more to make them large enough to serve subsequent phases that may (or may not) be built?

14.1.2 The analysis

The problem could be framed and explored using a Monte Carlo simulation (see page 92) and sensitivity analysis (see page 68), but analysis can start more simply with the project team's best guesses, followed by testing around those assumptions.

- What is the probability that the adaptability will be used? Rick feels that it is highly likely – perhaps an 80 per cent likelihood of it being used over the next thirty years. Others on the project team may feel it is more or less, but one has to start somewhere, and it can never be as much as 100 per cent – after all, the building may burn down first.

■   If it is used, when will it occur? Rick is reasonably sure it will not be used in the first five years, because of the time it would take any new breakthrough to be developed and come to market at a cost low enough for the client to buy in.

Rick knows that the initial extra costs will be about £35,000. The costs involved in completely reconstructing the roof some years in the future would be larger. Some rough calculations show a potential saving of about £200,000 if a solar installation is made and the roof is properly oriented. The potential is to make a considerable saving if a small 'toe hold' amount is spent.

This problem can be approached using a spreadsheet, setting out a series of years and the processes that occur in each. Using the spreadsheet's random number generator, such a model can calculate a present value of the savings based on if and when the adaptability is used. Combining these two factors will yield a probability that the adaptability will be used in any specific year. Through Monte Carlo simulation, multiple possible courses of the project can be projected and the present value of the adaptability explored.

A key question is what discount rate to use. From a social point of view, discounting should be undertaken at a low rate – perhaps the 3.5 per cent suggested in the Treasury Green Book. However, the investment is being made for an entrepreneur, so a higher private rate is probably more important to him. This underlines how such investments might be perceived differently, and why government subsidies are often needed to stimulate investment for the collective good.

A small piece of a simple simulation is shown in Figure 14.1. Essentially in year six, the first in which the installation of solar collecting equipment might be made, a random number generator is used to calculate the number of years until the solar installation occurs (when the benefit will be received). Another random process is used to determine whether or not that simulation run is included in the cases that actually use the adaptability. In the first series shown, the installation occurs relatively soon, and in the second case it occurs in the more distant future.

Figure 14.1 Adaptability for future solar collection: a section of the simulation												
Discount rate (real)		6.00%	Market discount rate									
Extra cost		35,000										
Probability of solar collection (ever)		90%										
Years before it might be used		5	(if less, decision-makers would probably know it and do it now)									
Years to end of option	30	25	(if beyond that, the building may have changed use, or roof may be reconstructed)									
Distribution function		4.50%	/year (after year 5)									
Performance difference between building with and without flexibility provisions			Maximum difference = £140,992									
*Average PV of savings =	68,326											
1st run												
Will flexibility be used?												
	Rand	0.43527	Prob of being used =		0.900							
Flexibility used/not used (1 = used)		1.00	Used									
With provisions	Use function:	<u>0</u>	<u>1</u>	<u>2</u>	<u>3</u>	<u>4</u>	<u>5</u>	<u>6</u>	<u>7</u>	<u>8</u>	<u>9</u>	
Rand								2	1	0	-1	
Savings from flexibility			0	0	0	0	0	0	0	200,000	0	
PV of savings	125,482		0	0	0	0	0	0	0	125,482	0	
2nd run												
Will flexibility be used?												
	Rand	0.44932	Prob of being used =		0.900							
Flexibility used/not used (1 = used)		1.00	Used									
With provisions	Use function:	<u>0</u>	<u>1</u>	<u>2</u>	<u>3</u>	<u>4</u>	<u>5</u>	<u>6</u>	<u>7</u>	<u>8</u>	<u>9</u>	
Rand								18	17	16	15	
Savings from flexibility			0	0	0	0	0	0	0	0	0	
PV of savings	49,396		0	0	0	0	0	0	0	0	0	

14.1.3 Results and their interpretation

The simulation was run 5,000 times, using the base data and assumptions. Figure 14.2 shows the distribution of outcomes. As set up by Rick, in 20 per cent of the cases the adaptability is not used, and the investment is lost. However, a decision to make the building adaptable is justified by the situations in which the benefits are received, even if they are at different times in the future, giving rise to a range of positive present values.

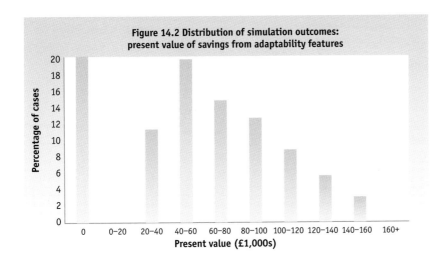

Figure 14.2 Distribution of simulation outcomes: present value of savings from adaptability features

The result indicates that the mean of the discounted benefits (the 'option value') was about £47,000, and exceeds the cost of doing the work, so proceeding with adapting the building for future solar energy collection is suggested. However, the limited difference suggests sensitivity testing around the assumptions, in particular the probabilities, which are best guesses. Alternatives might be tested around the uncertainties, and a chart developed. As a model has already been built, it is usually easy to change some of the input variables in a methodical way. Figure 14.3 shows an exploration of the 'likelihood of use' variable, suggesting that if the members of the project team believe that, all other things being equal (they rarely are), the probability of using the adaptability is 45 per cent or more (and the other assumptions are correct), so the investment is reasonable. Obviously, there is no way of confirming the probabilities, so the decision-makers will have to follow their own sense of what is likely. The decision still depends upon expert judgement, but is pushed back a stage so more specific and penetrating questions have to be answered.

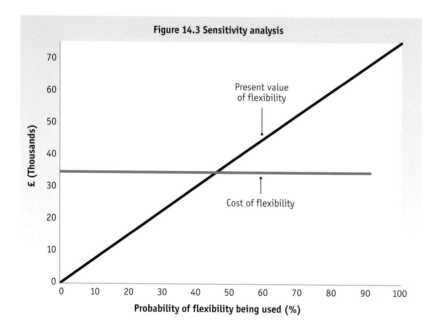

Figure 14.3 Sensitivity analysis

Whether or not flexibility options are ultimately exercised is uncertain, but project participants do have useful insights that can be used in making the decisions that have to be made. The default option, simply following the programme for the current use and situation, usually implies not including flexibility or adaptability, so forgoing any benefits. Exploration of decisions can give the members of a project team a better feel for the numerical characteristics of the decision. It cannot give an absolute answer, given that there is no possible way of knowing the actual probabilities that might be associated with this type of problem. Even with hindsight they cannot be known, because only one course of events occur, rather than the entire set of possible courses represented in a simulation.

How Does One Know if Real Options are Likely to be Important?

While financial options (and lease options) are explicitly created vehicles that usually come with the label 'option', real options often simply appear. Just identifying them can be difficult.

When should real options be considered?

- Uncertainty has to be present, and the greater the uncertainty the more likely it is that real options will be of consequence. If we could know the future, options would have no value. In the finance world the worth of a traded option relates to the volatility (a measure of uncertainty) of the underlying asset. Some buildings (or their components) exist in very stable circumstances, while for others uncertainty is rampant. That is why offices often have easily moved or demolished partitions, but housing is built using more permanent wall systems.
- The decision is not a now-or-never situation. As the project moves into the future, there has to be the possibility of decisions in the future, in particular about how the direction of the project might be changed to reflect new circumstances.
- The situation is irreversible. If you do not like the results, you cannot undo it and recover the resources used.

14.2 Real Options Can Help with Your Sustainability Shopping List

Real-options thinking can change the way a whole-life costing exercise might rank alternatives. Typically, in the design process, different alternatives are associated not only with different expected returns, but with different levels of flexibility. In the energy-saving choices used as an example in Chapter 4, the designer was considering more insulation, better windows or an improved mechanical system, each of which on its own offered the same return. In the real world resources are always limited, and the project may be restricted to only one of the alternatives.

Upgrading insulation tends to be messy, inconvenient and expensive – and it often involves re-cladding a building or rebuilding interior walls. It is almost a now-or-never decision. Insulation also has the advantage that it performs without the need for any maintenance by the occupant or owner. In those Victorian buildings that were insulated with rags or sawdust, the insulation continues to perform, though perhaps with some settlement or mouse holes. Upgrading windows is less of an inconvenience to building users, as it may be done when the building is not being used, perhaps at weekends or in the summer. Sometimes secondary glazing upgrades can be installed with no disruption at all. Mechanical systems do not have the long-term durability of windows or insulation, and elements of them are replaced periodically. While many buildings still have their Victorian windows, and use their Victorian heat distribution systems, very few are still using their original boilers. Boilers are easier to upgrade than distribution systems, and are replaced with newer technologies as they wear out or as replacement parts become unavailable.

In this case, it is easy to rank the three alternatives, because they offer the same initial yield. The best thing in such a case would be to select the one with the least flexibility for future upgrading – the enhanced insulation. Of course, a quantified analysis may be necessary for solving more complex problems.

While much real-options analysis can be mathematically complex, some problems can be solved without mathematics, simply by considering the characteristics of the alternatives, in particular associated flexibility.

14.3 Strategic Planning: Master Plans

The CABE document *Creating successful master plans: a guide for clients* (2004) states: 'The demand for strategic thinking about the process of urban and rural change is growing rapidly, as local authorities, regional development agencies, urban regeneration companies, housing market renewal pathfinders, private developers and communities alike need to think about physical change at a large scale.'[2]

Master plans are usually strategic outlines for multi-phase developments. They usually involve planning for sequential development of a series of project phases to be built as demand or funding materialises, so the time frame is often unknowable. Master plans may flow out of overall business plans, in which an organisation's values, missions and directions are stated. The creator of the property master plan may or may not have knowledge about the overall business plan, but if one exists a master plan should reflect it.

Perhaps the most important aspect is to avoid blocking future development opportunities. As some master plans cover long periods of time, sometimes a century or more, what form those development opportunities will take is very difficult or even impossible to predict.

Therefore a master plan should create new real options and safeguard existing ones. When a site is being developed in a highly uncertain context, perhaps to accommodate the needs of a high-tech company, concern about flexibility and alternative development options can add a considerable amount of value. Therefore a single forecast of what might be should not be the only basis for a master plan, but only one among many contributing elements. A housing development might be expected to last several decades with fewer changes – there is less inherent uncertainty in such a project.

This gives a different sense of what a development master plan should be. While initial stages may be described in some detail, with the reasonable expectation that it might actually be built, this is unlikely to be the case for subsequent phases. Rather than lay one or two schemes out in detail, it is more appropriate to explore and understand the alternative ways in which follow-on development might occur. In an options sense, a good master plan offers future decision-makers a variety of possibilities – they will exercise the options that they see as most appropriate at the time.

In Chapter 9, Mr Sullivan made provision for a future expansion to his phased shopping centre development. He might well consider how to use the remainder for offices, a hotel or housing. This might mean redesigning the circulation on the site now to achieve that flexibility. Of course, the less certain the environment, the more valuable those options will be. In a low-uncertainty context, little will change, so it is likely that it will always be better to build a shopping centre than offices; however, if there is ongoing turmoil in the local marketplace, it is possible that at some point offices or a hotel will be more attractive.

Master planning options do not always involve expansion. It is worth considering how a project can be trimmed back, perhaps by selling an unused part of a site (which means that access will be required). One office development studied by the authors did plan for contraction or even failure of the occupying organisation, by designing for possible subdivision for multi-tenant occupation. While few organisations want to talk about such things, failure of substantial businesses does occur, and while the business in that particular office building is still there, their multinational parent did in fact collapse.

In many cases, acquiring an option through a master plan costs nothing. However, in others there is some cost – perhaps a less convenient arrangement of the initial phases. In such cases, quantification may help one understand how much inconvenience to accept.

14.4 Real Options and the Decision-maker

There are many reasons why a basic understanding of real options and their implications will help the decision-maker, even if quantification is never attempted.

- Options proliferate in the property world. These include options to sell, develop, improve, subdivide, refurbish, change use and demolish. They are part of the rights that come with ownership. The building owner has more rights than the tenant. In the case of tenants, specific options are often delineated in the lease – options to expand, contract, renew or terminate. Some options come with labels, such as 'Option to Purchase', while others are very difficult to recognise. Some run with the property and are just there. Others may have to be acquired, such as an ability to easily subdivide space. Options can be destroyed as well as created. This is a big problem with the listing of buildings, because the option to redevelop the site is lost if the existing building is protected from demolition, and others, such as changing use, may be more expensive to exercise.

- The future is uncertain, so decisions made today may ultimately be ill-suited to future conditions. It might be better to create some options that can be used to respond to unfolding events.

When dealing with real options, quantification is sometimes demanded. While we know that flexibility in buildings is a good thing ('long life, loose fit'), usually flexibility comes at some cost. A column-free space is more flexible than space cluttered with columns, but the required long spans are usually more expensive. A quantitative analysis can help us answer the question about how much should be spent to get that flexibility. One of the authors recalls a major office landlord who had a 'no expansion option' policy with respect to leases. The reason was that if an expansion option was granted, the space over which it was granted would command a lower rent because the term would be limited in duration, because the space had to be recovered if the option was exercised. There were substantial costs to this policy in terms of good tenants who did not sign up because they needed the expansion options. In the absence of quantification, neither side really knew what was the best course of action.

Architects and managers are often required to create programmes or master plans. Yet this process may encourage concentrated thought about the details of the requirements as they exist in the present, without fully recognising the possibility of fundamental change and how the project under consideration might respond to unfolding conditions.

Architects (and sometimes other people too) often want to control everything, down to very small details, and indeed sometimes the details are important. But often it is best not to try to make all the decisions now. Decision-makers in the future will inevitably make decisions in ways that are appropriate to their own time and needs using the resources they have, and responding to information that may not now exist. The stock of capital is increased when we pass to future generations things that they can use.

14: Notes

1   Nichola Jones (2012), 'Turn a new leaf', *New Scientist*, 14 April, pp. 29–31
2   CABE – Commission for Architecture and the Built Environment (2004), *Creating Successful Masterplans: A Guide for Clients*, London: CABE, p. 9

And Finally ...

15

The world in which we live is totally obsessed with the concept of sustainability. The word is everywhere, but meanings can be elusive. Does it mean different things in different contexts? How does one balance the different elements of sustainability?

Clearly, moving towards more sustainable buildings is dependent upon making good decisions; poor decisions will lead to improper allocation of resources. But how can architects, developers, managers and others involved in creating buildings and infrastructure make sustainable decisions? It is not a simple task! Decision-making in architecture is fraught with complexity, contradictions, uncertainty, appalling data and very long time frames over which sustainability has to be addressed. Yet, even in the face of these difficulties, decisions do have to be made.

Decisions need to be made with great care and knowledge of many different aspects. It is important that decision-makers have a good understanding of how they define sustainability, in particular how they might balance economic, social/cultural and environmental concerns, because all are important to present and future well-being.

How do people in the built environment make decisions? It is inevitable that, given the myriad of decisions a busy practitioner might be making every day, many will be made by informed intuition. Yet an understanding of a set of basic tools that can be used to analyse decisions will help to improve the capability of the practitioner. A quick analysis can verify that the results are consistent with available data and reasonable assumptions. For major decisions, data gathering and numerical analysis may be in order.

A competent decision-maker should not only know about available tools, but needs to be aware of the biases that can creep in and affect decisions – both those that are part of our nature as humans, and those that are associated with group dynamics. The management literature is replete with studies of the factors that have led to major catastrophes.

The good news is that there are ongoing advances in decision-making practices. The field of real options has emerged over the past twenty years and is a natural enabler of sustainability in buildings. Computers, even just spreadsheets, allow the modelling of problems and the testing of solutions that would have been unfeasible a few decades ago. Without them, sensitivity testing and Monte Carlo analysis would be prohibitively time-consuming except for the largest and most consequential matters. We can anticipate further advances, and the appearance of useful dedicated software that will reduce time and effort and allow more ready access to databases.

Understanding the definitions of sustainability for the built environment, paying attention to our own strengths and weaknesses in decision-making, and remembering and using available and useful tools can better enable those involved to make not only adequate but truly exemplary sustainable decisions in building design.

Important Lessons:

- Sustainability as a concept needs to be understood by the decision-maker. Unless a decision-maker understands what he/she means by sustainability, making decisions to further it will be very difficult. Members of a project team may have different understandings of the concept, and the importance of the various components may vary depending upon the decision at hand.

- The key to sustainable decision-making is to avoid either overinvesting or underinvesting. Both represent a waste of resources that might have been used more effectively in some other way. Balance is important.

- Whole-life decision-making encompasses many tools. The decision-maker can use these at a range of levels: just the knowledge of them can help inform personal expert opinion, while some decisions will demand a comprehensive numerical analysis.

- There is rarely a 'right' decision. When uncertainty exists, the possibility of failure does too. Even with hindsight it is difficult to tell whether the best decision was made, or whether the results were simply a matter of luck. What one strives for is some sort of 'optimum' decision, often one that is robust and so capable of yielding good outcomes over a range of possible future conditions.

- Decision-making is more than collecting data and putting it into analytical engines. Understanding the situation is more important – especially the nature of the uncertainties surrounding and influencing it. For this reason, a pencil-and-paper analysis, perhaps drawing a decision tree and using a few rough numbers can be quite effective.

- You don't have to make all of the decisions now. People in the future will have their own needs and expectations, as well as access to more information, and sometimes the best thing you can do is recognise the situation and provide them with the flexibility to make their own decisions.

- Becoming an expert and effective decision-maker is a personal journey and requires ongoing effort as new techniques and computer-based methods become available.

Bibliography

Adams, Tim (2012), 'This much I know: Daniel Kahneman', *Guardian*, 8 April, 2012, http://www.guardian.co.uk/science/2012/jul/08/this-much-i-know-daniel-kahneman, accessed 11 September, 2012.

Addis, Bill (2007), *Building: 3000 Years of Design, Engineering and Construction*, London: Phaidon Press.

Appelt, Kirstin C.; Milch, Kerry F.; Handgraaf, Michel J.J. and Weber, Elke U. (2011), 'The Decision Making Individual Differences Inventory and guidelines for the study of individual differences in judgment and decision-making research', *Judgment and Decision Making*, 6(3), pp. 252–62.

Ashuri, Baabak; Lu, Jian and Kashani, Hamed (2011), 'A real options framework to evaluate investments in toll road projects delivered under the two-phase development strategy', *Built Environment Project and Asset Management*, 1(1), pp. 14–31.

Assaf, Sadi et al. (2002), 'Assessment of the problems of application of life cycle costing in construction projects', *Cost Engineering*, 44(2).

Athena Sustainable Materials Institute and Morrison Hershfield Limited (2009), *A Life Cycle Assessment Study of Embodied Effects for Existing Historic Buildings*, www.athenasme.org/publications/docs/Athena_LCA_for_Existing_Historic_Buildings.pdf, accessed 20 June, 2011.

Atkinson, Giles and Mourato, Susana (2008), 'Environmental cost–benefit analysis', *Annual Review of Environmental Resources*, 33(3), pp 17–44.

Ayres, Robert U.; van den Bergh, Jeroen C.J.M. and Gowdy, John M. (2001), 'Strong versus week sustainability: Economics, Natural Sciences, and Consilience', *Environmental Ethics*, 23(2), pp. 155–68.

Baron, Jonathan (2008), *Thinking and Deciding* (4th Edition), Cambridge: Cambridge University Press.

Bordass, Bill (2000), 'Cost and value: fact and fiction', *Building Research and Information*, 28(5/6), pp. 338–52.

Boer, F. Peter (2002), *The Real Options Solution: Finding Total Value in a High-Risk World*, New York: John Wiley & Sons, Inc.

BRE (2000), 'Assessing environmental impacts of construction (BRE Digest 446)', Watford: Building Research Establishment.

British Standards Institute (1998), *Guide to the Principles of the Conservation of Historic Buildings BS 7913:1998*.

BSI Group (2011), 'Specification for the assessment of the life cycle greenhouse gas emissions of goods and services, PAS (Publicly Available Specification) 2050:2008', http://www.bsigroup.com/upload/Standards%20&%20Publications/Energy/PAS2050.pdf, accessed 24 January, 2012.

Bullis, Kevin (2009), 'Igniting fusion', *Technology Review*, 112(4), pp. 24–31.

CABE – Commission for Architecture and the Built Environment (2004), *Creating Successful Masterplans: A Guide for Clients*, London: CABE.

Cameron, Silver Donald (2010), 'Hedonic Indicators: Bhutan takes the next step in systematizing happiness', *The Walrus*, 7(3), pp. 19–21.

Cambridge Modern History, Volume III, The Wars of Religion, Chapter XIV, The End of the Italian Renaissance, www.third-millennium-library.com/reading/MODERN-HISTORY/Wars_of_Religion/14-END-OF-THE-ITALIAN-RENAISSANCE.htm, accessed 2 May, 2011.

Carver, C. S.; Scheier, M. F. and Segerstrom, S. C. (2010), 'Optimism', *Clinical Psychology Review* 3, pp. 879–89.

Chinneck, John W. (2000), 'Practical Optimization: a Gentle Introduction', www.sce.carleton.ca/faculty/chinneck/po.html, accessed 10 May, 2011.

Clift, Mike and Bourke, Kathryn (1999), *Study on Whole Life Costing, Report Number CR366/98*, Watford, Building Research Establishment.

Cole, R.J. (2003), 'Green buildings – reconciling technological change and occupant expectations' in Cole, R.J. and Lorch, R. (eds) (2003), *Buildings, Culture and Environment*, Oxford: Blackwell.

Copeland, Tom and Antikarov, Vladimir (2001) *Real Options: A Practitioner's Guide*, New York: Texere.

Cottam, David (1986), *Sir Owen Williams, 1890–1969*, London: Architectural Association.

Davis Langdon Management Consulting (2007), 'Life Cycle Costing (LCC) as a contribution to sustainable construction: a common methodology', http://ec.europa.eu/enterprise/sectors/construction/files/compet/life_cycle_costing/final_report_en.pdf, accessed 12 April, 2012.

de Neufville, Richard and Scholtes, Stefan (2011), *Flexibility in Engineering Design*, Cambridge, Mass: MIT Press.

Department of for Environment, Food and Rural Affairs (2007), *The Social Cost of Carbon and the Shadow Price of Carbon: What They Are, And How to Use Them in Economic Appraisal in the UK*, London: Economics Group, Defra.

Dijksterhuis, Ap and Nordgren, Loran F. (2006), 'A theory of unconscious thought', *Perspectives on Psychological Science*, 1, pp. 95–109.

Dohmen, Thomas; Falk, Armin; Huffman, David and Sunde, Uwe (2007), 'Are risk aversion and impatience related to cognitive ability?', *Discussion Paper IZA DP* no. 2735, Bonn: Institute for the Study of Labor, http://ftp.iza.org/dp2735.pdf, accessed 15 December, 2011.

Drummond, Helga (2012), *Guide to Decision Making: Getting it more right than wrong*, Hoboken: Wiley.

Durand, Jean-Nicolas-Louis (2000), *Précis of the Lectures on Architecture*, Los Angeles: Getty Foundation (first published in Paris, 1821).

Edwards, Brian (2005), *Rough Guide to Sustainability* (2nd Edition), London: RIBA Enterprises.

Ellingham, Ian and Fawcett, William (2006) *New Generation Whole-Life Costing: Property and Construction Desision-Making Under Uncertainty*, London: Taylor and Francis.

English Heritage (2008), *Conservation Principles, Policies and Guidance for the Sustainable Management of the Historic Environment*, London: English Heritage.

Ferguson, Niall (2009), *The Ascent of Money: A Financial History of the World*, New York: Penguin.

Fisk, David J. (1976), *Energy Conservation: Energy Costs and Option Value*, Watford: Building Research Establishment.

Fisk, D.J. and Kerherve, J. (2006), 'Complexity as a cause of unsustainability', *Ecological Complexity*, 3, pp. 336–43.

Flyvbjerg, Bent (2008), 'Curbing optimism bias and strategic misrepresentation in plannning: reference class forecasting in practice', *European Planning Studies*, 16(1).

Food and Agriculture Organization of the United Nations (2006), *Undernourishment Around the World: Counting the Hungry: Trends in the developing world and countries in transition*, Rome: FAO.

Frowen, Stephen F. (1990), *Unknowledge and Choice in Economics, Proceedings of a conference in honour of G.L.S. Shackle*, Basingstoke, Macmillan.

Gardner, Dan (2008), *Risk*, Toronto: McClelland & Stewart.

Ghafele, Roya, 'Getting a Grip on Accounting and Intellectual Property', http://www.wipo.int/sme/en/documents/ip_accounting.html, accessed 21 October, 2011.

Gollier, Christian (2001), *The Economics of Risk and Time*, Cambridge, Mass.: MIT Press.

Guo, Jiehan; Hepburn, Cameron J. and Tol, Richard S.J. (2006), 'Discounting and the social cost of carbon: a closer look at uncertainty', *Environmental Science and Polity*, 9, pp. 205–16.

Harvard Business Review (2011), *Making Smart Decisions*, Boston: Harvard Business Review Press.

Hassett, Kevin A. and Metcalf, Gilbert E. (1993), 'Energy conservation investment: Do consumers discount the future correctly?', *Energy Policy*, 21(6), pp. 710–16.

Hauff, Volker (2007), 'Brundtland Report: A 20 Years Update, Keynote speech, European Sustainability', Berlin 2007, EXB07, http://www.nachhaltigkeitsrat.de/uploads/media/ESB07_Keynote_speech_Hauff_07-06-04_01.pdf, accessed 3 March, 2012.

Hawken, Paul; Lovins, Amory and Lovins, L. Hunter (1999), *Natural Capitalism*, Boston: Little, Brown and Company.

Hayes, John R. (1981), *The Complete Problem Solver*, Philadelphia: The Franklin Institute.

Hepburn, Cameron (2007), 'Valuing the far-off future: discounting and its alternatives', from Atkinson, Giles; Dietz, Simon and Neumayer, Eric (eds), *Handbook of Sustainable Development*, Cheltenham: Edward Elgar.

Hines, Mary Alice (2001), *Japan Real Estate Investment*, Westport, Conn: Quorum Books.

HM Treasury (2003), *The Green Book: Appraisal and Evaluation in Central Government*, London: HM Treasury.

HM Treasury 'Supplementary Green Book Guidance: Optimism Bias', http://www.hm-treasury.gov.uk/d/5(3).pdf, accessed 9 March, 2012.

Innes, Harold A. (1951), *The Bias of Communication*, Toronto: University of Toronto Press.

Jacobs, Jane (1961/1963), *The Death and Life of Great American Cities*, New York: Random House.

Jones, Nichola (2012), 'Turn a new leaf', *New Scientist*, 14 April, pp. 29–31,

Jordaan, Ian (2005), *Decisions under Uncertainty*, Cambridge: Cambridge University Press.

Kahneman, Daniel and Frederick, Shane (1990), 'Experimental tests of the endowment effect and the coase theorem', *Journal of Political Economy*, 98, pp. 1325–48.

Kahneman, Daniel; Slovic, Paul and Tversky, Amos (eds) (1985), *Judgement under Uncertainty: Heuristics and biases*, Cambridge: Cambridge University Press.

Keynes, J.M. (1933), *Essays in Biography*, London: Palgrave Macmillan.

Keynes, J.M. (1936, 1976 reprint), *The General Theory of Employment, Interest and Money*, London: Macmillan.

Knight, Frank H., (1921), *Risk, Uncertainty and Profit*, Boston: Houghton Mifflin (reprinted 1971 Chicago: University of Chicago Press).

Knox, Paul and Mayer, Heike (2009), *Small Town Sustainability*, Basel: Birkhauser.

Kossinets, Gueorgi; and Watts, Duncan J. (2006), 'Empirical analysis of an evolving social network', *Science*, 6 January, 311(5757), pp. 88–90.

Landes, David S. (1998), *The Wealth and Poverty of Nations*, New York: Norton.

Lichtenstein, S; Slovic, P.; Fischhoff, B.; Layman. M. and Combs, B. (1978), 'Judged frequency of lethal events', *Journal of Experimental Psychology: Human Learning and Memory*, 4(565), from Baron p. 138.

Liddell, Howard (2008), *Eco-Minimialism: The Antidote to Eco-Bling*, London: RIBA Publishing.

Loasby, Brian (1990), 'The use of scenarios in business planning', from Frowen, Stephen F. (ed), *Unknowledge and Choice in Economics, Proceedings of a conference in honour of G.L.S. Shackle*, Basingstoke: Macmillan.

Loasby, Brian (2010), 'Uncertainty and Imagination, Illusion and Order: Shackleian Connections' the G.L.S. Shackle Biennial Memorial Lecture, 4 March, St Edmund's College, Cambridge.

Lord, Charles G.; Ross, Lee and Lepper, Mark R. (1979), 'Biased assimilation and attitude polarization: the effects of prior theories on subsequently considered evidence', *Journal of Personality and Social Psychology*, 37(11), pp. 2098–109.

Lovallo, Dan and Kahneman, Daniel (2011), 'Delusions of success: how optimism undermines executives' decisions', pp. 29–49, in *Harvard Business Review: Making Smart Decisions*, Boston: Harvard Business Review Press.

Lutzkendorf, Thomas and Lorenz, David (2005), 'Sustainable property investment: valuing sustainable buildings through property performance assessment', *Building Research & Information*, 33(3), pp. 212–34.

Milne, Glen (2011A), 'Foresight: Understanding the future security environment', *Vanguard*, www.vanguardcanada.com/TheBenefitOfForesightMilne, accessed 15 April, 2011.

Milne, Glen (2011B), 'The Internal War on Innovation', Canadian Government Executive, 4 March, 2011.

Mithraratne, Nalanie; Vale, Brenda and Vale, Robert (2007), *Sustainable Living: the Role of Whole Life Costs and Values*, Oxford: Elsevier.

Moor, Steven A. (ed) (2010), *Pragmatic Sustainability: Theoretical and Practical Tools*, London: Routledge.

Myers, A.R. (1975), *Parliaments and Estates in Europe to 1789*, London: Thames and Hudson.

Obasli, Aylin (2008), *Architectural Conservation*, Oxford: Blackwell Science.

OECD (1998), *Human Capital Investment: An International Comparison*, Paris: Organisation for Economic Cooperation and Development.

OGC (2009), 'Life Cycle Costing', www.ogc.gov.uk/implementing_plans_introduction_life_cycle_costing_.asp, accessed 30 June, 2011.

Onis, Mercedes de and Blossner, Monika (2003), 'The World Health Organization Global Database on child growth and malnutrition: Methodology and applications', *International Journal of Epidemiology*, 32, pp. 518–26.

Pearce, D.W. and Nash, C.A. (1981, 1991 reprint), *A Social Appraisal of Projects: A Text in Cost-Benefit Analysis*, Basingstoke and London: Macmillan.

Pearce, David and Barbier, Edward B. (2000), *Blueprint for a Sustainable Economy*, London: Earthscan.

Pearce, David (2001), 'A bright green', *New Scientist*, 171(2309), 22 Sept, p. 50.

Pollard, Doug (2009), 'Changing into tails', *OAA Perspectives*, Spring, pp. 11–12.

Pope, Stephen (2006), 'Green green everywhere, but not a moment to think, or not there yet', *OAA Perspectives*, Fall, 24(3).

Power, Anne (2008), 'Does demolition or refurbishment of old and inefficient homes help to increase our environmental, social and economic viability?', *Energy Policy*, 36, pp. 4487–501.

Price, Richard; Thornton, Simeon and Nelson, Stephen (2007), 'The Social Cost of Carbon and the Shadow Price of Carbon: What they are, and how to use them in economic appraisal in the UK', Department for Environment, Food and Rural Affairs.

Quinn, Robert E. (1988), *Beyond Rational Management*, Oxford: Jossey-Bass.

Rosenhead, Jonathan; Elton, Martin and Gupta, Shiv (1972), 'Robustness and optimality as criteria for strategic decisions', *Operations Research Quarterly*, 23(4), pp. 413–31.

Ross, Barbara (2009), 'Teaching Architecture Students to Think Critically about Sustainable Design', unpublished paper: Centre for Teaching Excellence, University of Waterloo.

Ross, Barbara (2009), 'Design with Energy in Mind', unpublished master's dissertation, University of Waterloo.

Royal Institute of British Architects (2009), *Climate Change Toolkit 08 – Whole Life Assessment for Low Carbon Design*, London: RIBA, http://www.architecture.com/FindOutAbout/Sustainabilityandclimatechange/ClimateChange/Toolkits.aspx, accessed 23 January, 2012.

Russell, Clifford S. (2001), *Applying Economics to the Environment*, Oxford: Oxford University Press.

Salganick, Matthew; Dodds, Peter and Watts, Duncan (2006), 'Experimental study of inequality and unpredictability in an artificial cultural market', *Science*, 311, 10 February, pp. 854–6 and supporting material on *Science* website.

Samuelson, Paul (1967), *Economics* (7th edition), New York: McGraw-Hill.

Sanstad, Alan H.; Blumstein, Carl and Stoft, Steven E. (1995), 'How high are option values in energy-efficiency investments?' *Energy Policy*, 23(9), pp. 739–43.

Saynajoki, Antti (2011), 'An empirical evaluation of the significance of the carbon spike in the construction industry', proceedings of the World Sustainable Building Conference, 18–21 October, Helsinki.

Schon, Donald A. (1987), *Educating the Reflective Practitioner*, San Francisco: Jossey-Bass.

Seale, Darryl A.; Rapopart, Amnon and Budescu, David B. (1995), 'Decision-making under strict uncertainty: an experimental test of competitive criteria', *Organizational Behavior and Human Decision Processes*, 64(1), October, pp. 65–75.

Shackle, G.L.S. (1961), *Decision, Order and Time in Human Affairs*, Cambridge: Cambridge University Press.

Schank, Roger C.; Lyras, Dimitris and Soloway, Elliot (2010), *The Future of Decision Making: How Revolutionary Software Can Improve the Ability to Decide*, New York: Palgrave Macmillan.

Sharot, Tali (2011), *The Optimism Bias: A Tour of the Irrationally Positive Brain*, New York: Pantheon.

Shermer, Michael (2009), 'Agenticity', *Scientific American*, June, p.36.

Sirgy, J. Joseph (1982), 'Self-concept in consumer behavior: a critical review', *Journal of Consumer Research*, 9, pp. 287–300.

Smith, Laurence C. (2010), *The World in 2050: Four forces shaping civilization's northern future*, Dutton, New York.

Solomon, Lawrence (2009), 'Danger overhead', *National Post*, 18 April, p. FP15.

Spence, Rick (2009), 'No crystal ball? Consider using a matrix', *National Post*, 22 June, p. FP4.

'Standardized Method of Life Cycle Costing for Construction Procurement: A supplement to BS ISO 15686-5 Buildings & constructed assets – Service life planning – Part 5: Life cycle costing' (2008), BSI Technical Subcommittee B/500/3, Durability, London: BSI.

Steele, James (1997), *Sustainable Architecture*, New York: McGraw-Hill.

Stern, Nicholas (2006), *Stern Review on the Economics of Climate Change*, London: HM Treasury, http://webarchive.nationalarchives.gov.uk/+/http:/www.hm-treasury.gov.uk/sternreview_index.htm, accessed 15 September, 2011.

Thornton, Mark (2007), 'Richard Cantillon and the discovery of opportunity cost', *History of Political Economy*, 39(1), pp. 97–119.

Taleb, Nassim Nicholas (2001), *Fooled by Randomness*, London: Texere.

Torbert, W.R. (1987), *Managing the Corporate Dream: Restructuring for long-term success*, Homewood, Ill: Dow-Jones Irwin.

Trevelyan, Rose (2008), 'Optimism, overconfidence and entrepreneurial activity', *Management Decision*, 46(7), pp. 986–1001.

Turner, Kerry (2005), 'The "Blueprint" Legacy – a review of Professor David Pearce's contribution to environmental economics and policy', CSERGE Working Paper PA 05–01, http://cserge.uea.ac.uk/sites/default/files/pa_2005_01.pdf, accessed 23 October, 2011.

Veefkind, Menno (1998), 'Life cycle costing for green product development', Technical University of Delft paper (from the T.U.Delft website).

Walker, Peter and Greenwood, David (2002), *Risk and Value Management*, London: RIBA Enterprises.

Wallsgrove, John (2008), 'The justice estate's energy use', *Context*, 102, http://www.ihbc.org.uk/context_archive/103/wallsgrove/page.html, accessed 12 August, 2011.

Warrick, Chris (2012), 'Fusion turns to engineering', *Ingenia*, 52, pp. 39–43.

Webb, Richard (2008), 'Online shopping and the Harry Potter effect', *New Scientist*, 22 December, 2687.

Williams, Owen (1927), 'Factories', *RIBA Journal*, 16 November, pp. 54–5.

World Commission on Environment and Development (1987), *Our Common Future (Brundtland Report)*, Oxford: Oxford University Press.

Yudelson, Jerry (2009), *Green Building Trends: Europe*, Washington: Island Press.

Vyas, Ujjval (2007), 'An unsustainable policy: ignoring the difficult questions surrounding sustainable building is a dangerous path', *OAA Perspectives*, Winter 2007, pp. 39–40.

Vyas, Ujjval (2009), *Sustainable Design and Construction: When Green Turns Red*, Chicago: American Bar Association.

Wals, Arjen, E.J. and Jickling, Bob (2002), '"Sustainability" in higher education: from doublethink and newspeak to critical thinking and meaningful learning', *International Journal of Sustainability in Higher Education*, 3(3), pp. 221–32.

Webb, Richard (2008), 'Online shopping and the Harry Potter effect', *New Scientist*, 2687, 20/27 December, pp. 52–5.

Weybrecht, Giselle (2010), *The Sustainable MBA: The Manager's Guide to Green Business*, Chichester: Wiley.

Woolley, Simon (2009), *Sources of Value: A practical guide to the art and science of valuation*, Cambridge: Cambridge University Press.

Yoon, K. Paul and Hwang, Ching-Lai (1995), *Multiple Attribute Decision-Making*, Thousand Oaks, California: Sage.

Zeckhauser, R (1975), 'Procedures for Valuing Lives', *Public Policy*, 23, 419–6

Index

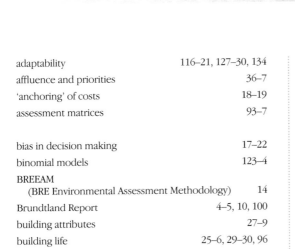